PRAISE FOR *NEUROSCIENCE FOR LEADERS*

'There are many books on leadership – perhaps too many! However, this book is well worth reading. The authors have done a really great job in bringing together developments in neuroscience and thinking about leadership to produce a thought provoking view of leadership. They make the complex area of neuroscience very accessible and use this to produce a really helpful framework for thinking about leadership. The authors have achieved a fine balance between the rigour of their arguments and the practical applications of the framework that they have developed. Even if you have a lot of books on leadership on your shelves you should make room for this one.'
Malcolm Higgs, Professor of Organization Behaviour and HRM, Southampton Business School, University of Southampton

'Neuroscience is invading every scientific, economic and social aspect more and more. This book is the brilliant confirmation that understanding the mechanisms, processes and functions of the brain can help us solve problems, distinguish leaders from regular managers and uncover the right answers for generating growth.'
Matteo Venerucci, Cognitive Psychologist and Brain Propaganda Founder

'I firmly believe that business results are the direct outcome of collective efforts, guided by a strong vision and even stronger leadership. However, until now the true nature of leadership was difficult to explain, regardless experiencing it daily. Dr Dimitriadis and Dr Psychogios have managed to untangle the mystery of effective leadership by revealing how the brain works in complex business environments. Read it, practise it and become a great leader!'
Maria Anargyrou-Nikolic, General Manager, Coca-Cola HBC Slovenia, Croatia, Bosnia and Herzegovina

'This book makes the complex subject of neuroleadership easy to understand. It brings to the table simple yet effective recommendations on how to make better decisions to avoid the trap of human bias. It is anchored in solid science so it will make even the most sceptical reader revisit their own biases. Highly recommended.'
Patrick Renvoise, co-author of *Neuromarketing: Understanding the buy button in your customer's brain* and Chief Neuromarketing Officer, SalesBrain

'Business is about people, not numbers. People make the numbers by innovating internally and delivering new solutions and experiences externally. This book is the starting point for anyone who wants to discover how our brain drives behaviour and makes leaders succeed. Highly recommended.'
Vojislav Lazarević, Executive Board Chairman and General Manager, Piraeus Bank AD Belgrade

'An amazing compendium that will certainly assist managers and professionals in advancing their leadership skills. The book deals with the subject of this interdisciplinary research to resolve complex problems while facing challenging environments. The authors have included numerous examples from a practical and holistic perspective in order to illustrate how to create common ground about our adaptive leadership and how to put into action our leadership brain capabilities by applying their proposed BAL model. It is a fascinating book which will change your life.'
Panayiotis H Ketikidis BSc MSc PhD, Professor, University of Sheffield International Faculty, City College and Chairman of the South East European Research Centre

'A must-have for all who believe that leadership is both a brand differentiator and a business driver, and are eager to engage themselves and others in enhanced performance.'
Hajnalka Stavinovszky, Marketing Competence Development Specialist, IKEA Retail Services AB

'With leadership remaining an ongoing challenge, this book finally sets the record straight that intelligence matters not least because it combines cognitive, emotional and intuitive dimensions.'
Elena Antonacopoulou, Professor, GNOSIS, University of Liverpool Management School

Neuroscience for Leaders

Neuroscience for Leaders
A brain-adaptive leadership approach

Nikolaos Dimitriadis and
Alexandros Psychogios

LONDON PHILADELPHIA NEW DELHI

Publisher's note

Every possible effort has been made to ensure that the information contained in this book is accurate at the time of going to press, and the publishers and authors cannot accept responsibility for any errors or omissions, however caused. No responsibility for loss or damage occasioned to any person acting, or refraining from action, as a result of the material in this publication can be accepted by the editor, the publisher or the authors.

First published in Great Britain and the United States in 2016 by Kogan Page Limited

Apart from any fair dealing for the purposes of research or private study, or criticism or review, as permitted under the Copyright, Designs and Patents Act 1988, this publication may only be reproduced, stored or transmitted, in any form or by any means, with the prior permission in writing of the publishers, or in the case of reprographic reproduction in accordance with the terms and licences issued by the CLA. Enquiries concerning reproduction outside these terms should be sent to the publishers at the undermentioned addresses:

2nd Floor, 45 Gee Street	1518 Walnut Street, Suite 1100	4737/23 Ansari Road
London	Philadelphia PA 19102	Daryaganj
EC1V 3RS	USA	New Delhi 110002
United Kingdom		India

© Nikolaos Dimitriadis and Alexandros Psychogios 2016

The right of Nikolaos Dimitriadis and Alexandros Psychogios to be identified as the authors of this work has been asserted by them in accordance with the Copyright, Designs and Patents Act 1988.

ISBN   978 0 7494 7551 2
E-ISBN   978 0 7494 7552 9

British Library Cataloguing-in-Publication Data

A CIP record for this book is available from the British Library.

Library of Congress Control Number

2016934154

Typeset by Graphicraft Limited, Hong Kong
Print production managed by Jellyfish
Printed and bound in Great Britain by CPI Group (UK) Ltd, Croydon CR0 4YY

CONTENTS

Acknowledgements x

Introduction: The leadership enigma and the human brain 1

The BAL approach in brief 8
Keep in mind while reading the book 11
References 13

PILLAR 1 Thinking 15

01 Powerful brain, powerful leader 17

The energy-devouring brain 18
The willpower muscle for leaders 19
Higher-level thinking 21
Strong values 21
Feedback 23
Burnout syndrome 23
Multitasking yourself into cognitive load 26
Keep in mind 30
References 31

02 Clear mind, strong direction 33

The evolution of the survival-obsessed brain 34
Beware of patterns bearing gifts 36
Together we stand... or do we? 37
Brain misfires 39
Corporate cultures favouring the biased mind 39
Asking is winning 41
Enquiring with style 43
When the ape takes control 46
Keep in mind 48
References 49

03 Higher performance, more followers 51

The ever-changing brain 52
Purpose above all 54
Flow to greatness 59
Creativity killed the competition 62
If memory serves me right 67
Adapt, bet and grow 70
Keep in mind 74
References 74

Summary of Pillar 1 76

PILLAR 2 Emotions 79

04 More emotion, better decisions 81

The emotion-run brain 83
Emotional style 85
From mood to great feelings 94
EQ as an empowerment qualification 97
Keep in mind 100
References 100

05 Right emotion, right action 103

The basic emotions in the brain 105
Elementary Dr Plutchik! 111
Bliss leadership 116
Connecting the emotional dots 121
Keep in mind 124
References 124

Summary of Pillar 2 127

PILLAR 3 Brain automations 129

06 Gut reaction, faster solution 131

The mind-controlling brain 133
It's prime time 136
New habits, old habits 143

Let's get physical   148
Expertise and automaticity   152
Keep in mind   155
References   155

Summary of Pillar 3   160

PILLAR 4   Relations   161

07   **More connected, more successful**   163

The socially wired brain   169
I know what you think   172
The mirrors in our brains   177
Human connectivity   178
Trust the amygdala and your brain chemicals   186
Keep in mind   189
References   189

08   **Brain communication, better persuasion**   193

Persuading the brain to act   194
Cialdini's influence   200
Talk to the brain   205
Persuasion stimuli   210
Keep in mind   213
References   213

Summary of Pillar 4   216

Concluding remarks: The future, the brain and the BAL approach   219

Keep in mind   225
References   225

Epilogue   227
Index   229

ACKNOWLEDGEMENTS

We would like to thank the dedicated, experienced and professional team at Kogan Page for its invaluable support.

We also want to thank all our students, colleagues, corporate clients and audiences for their enthusiastic reception of the ideas of this book and for their constant provision of fresh challenges, perspectives and lessons. Our brains keep flourishing because of you!

Most of all, we thank our families for their unlimited understanding and love. This book is dedicated to them.

Introduction
The leadership enigma and the human brain

W*hat makes a great leader?* This profoundly important question has been on the minds of scientists, philosophers, professionals, and even ordinary citizens, around the globe for centuries. It was also the main question on the front cover of the November 2015 issue of the *Harvard Business Review*, a widely influential magazine for management, showcasing the obsession of business and organizations with leadership. Regardless of its importance, though, the answer to this question is perpetually elusive. Many theories and models have been proposed in answer to it but none has succeeded in capturing leadership in its entirety. Until now. We believe that the leadership enigma finally has a convincing answer. An answer rooted in a specific and very tangible source: the brain.

Leadership is not an exact science. It is not a pure art either. It is not even a set of separate thoughts, feelings or actions. Leadership is an attitude. And like all attitudes in classical psychology theory, it simultaneously consists of thoughts, feelings and behaviours (Rosenberg and Hovland, 1960) in a sum greater than its parts. As an attitude, leadership can be found anywhere and everywhere in everyday life. It is not defined by hierarchies. It is not a given qualification. We often like to view leadership as a wild flower. Wild flowers can grow in many places as long as there are appropriate conditions. Similarly, leadership can emerge if it finds appropriate conditions within and outside organizations.

As a significant social phenomenon, leadership has been extensively studied. The main target in investigating leadership was, and still is, to identify the aforementioned conditions that influence it. Accumulative knowledge on leadership has been formulated by diverse sciences such as biology, psychology, sociology, political science, anthropology and history, among others. In organizational science, the critical impact of engaging leadership behaviours on success has been well established (Higgs and Rowland, 2011). But, as mentioned above, regardless of the amount of knowledge we currently have about leadership, we are still far from understanding the phenomenon holistically and even further from learning how to develop an effective leadership attitude. The reason for this is simple. Leadership is not a static phenomenon. It is a dynamic one that constantly evolves. We claim that to understand leadership, as a dynamic phenomenon, we need to take a dynamic approach. In other words, we need to always expand our ability to understand leadership as an attitude based on new forms of emerging knowledge. In this direction, the real breakthrough today is neuroscience.

Over the last 20 years, leaps in technology have helped neuroscientists study and understand the human brain better than ever before in the whole of human history. Ground-breaking insights started to emerge regarding neuroanatomy, synaptic development and brain functioning that significantly influenced the way we understand, not just the brain's inner workings, but, more profoundly, our individual and social attitudes. As a consequence, neuroscience has helped increase our understanding of leadership as an attitude within organizations. In fact, although neuroscience has had a more immediate impact on other business-related fields, such as marketing and communications, attention is visibly focusing on its effects on leadership, or on what Rock and Ringleb (2009) call *neuroleadership*. These two have defined neuroleadership as the study of the biological micro-foundations of the interpersonal, influence-based relationships among leaders and their followers. Neuroleadership claims that by understanding how the human brain works, leaders can have a real advantage in engaging themselves and others in enhanced performance (Rock and Tang, 2009). In support of this view, Henson and Rossou (2013) argued that there is already evidence that leaders become more effective when stimulating themselves and their teams to make better use of their brains' capacities. This is done mainly by enhancing individual brain strength, cultivating healthy relationships and developing high-quality collective thinking (Henson and Rossou, 2013). There is an increasing number of voices suggesting that neuroscience opens new paths for leadership understanding and for modifying the all-important leadership attitude.

What seems to be less studied and developed though are practical approaches of making leaders aware of their brain flexibility and capacity, as well as of how exactly these can influence leadership attitudes in the real world. Although there is a relatively large number of publications on the subject, very few provide a step-by-step, concrete approach to harnessing this new knowledge in a simple, systematic and practical way for the professionals that crave it the most: business and other organizational managers.

The aim of this book is to offer a practical and holistic approach to understanding and implementing the leadership brain. Drawing on the growing knowledge on the brain in neuroscience and behavioural sciences, as well as from traditional leadership and business thinking, this book uncovers feasible ways to boost your leadership brain. This is crucial for challenging business and organizational environments. By using diverse and cutting-edge global research we build a comprehensive, concrete but, at the same time, simple to use approach to help managers and professionals improve their leadership capabilities. We propose the brain adaptive leadership (BAL) approach as a way of thinking, feeling and acting within organized social entities.

Brain adaptive leadership is an attitudinal approach that individuals can follow in their attempt to recalibrate their brains and mould their behaviour accordingly to lead projects, processes and people. This approach falls under the wider field of applied neuroscience. The accent is on the 'applied', suggesting that this book is for those who need to implement new and effective leadership approaches as quickly as possible simply because existing ones are not performing particularly well.

The BAL approach consists of three core elements: brain; adaptive; leadership.

The brain

We live in the new century of the brain. This is what the renowned magazine *Scientific American* declared on the cover of its March 2014 issue. Over the last 20 years neuroscience, behavioural economics and other scientific disciplines have shattered many myths about decision making and human behaviour, especially the stubborn fixation on the notion that logic rules. This belief is fast going out of fashion, since our capacity for rational and analytical thinking is found not to be *the* absolute human capability. Important? Yes. The most important? No. Neuroscience, as we discuss in detail in the book, is clear: humans are not, and should not be, purely rational creatures. Replacing rationality with a more complex, deep and, dare we say, holistic view of the inner workings of the human brain and of how it influences our lives, is crucial for improving human behaviour. Consequently, over the last decade or so, there have been numerous studies and books on neuroscience, the brain and the hidden forces shaping behaviour. At the same time, there is a growing interest from people in various professions and fields to learn these new, brain-based rules of human behaviour.

Advancements in medical technology proved to be extremely helpful in this area. Imaging technologies (such as functional Magnetic Resonance Imaging – fMRI) and other neural, biometric and brainwave-capturing technologies have contributed dramatically to the way the brain is studied (Gorgiev and Dimitriadis, 2015). The key outcome of this is that many new insights about neural processes, development and connectivity have been uncovered. These findings influence not only the way in which we understand brain function per se. More importantly, they have a profound impact on revealing the ways in which brain functions influence our individual and collective decision making, and consequently, personal and social behaviour. In other words, these developments are drastically developing our understanding of social phenomena, such as leadership and change, in organizations.

For us, it is all in the brain. Leadership research and practice should start and end with the brain: on a single brain as much as on collective brains, within and outside organizations. Discussing leadership without neuroscience – and also psychology, anthropology and behavioural economics – is not only inadequate but can be seriously misleading. Our thought patterns, analytical skills, moods, emotional reactions, habits, relation-building and communication skills, our ability to change fast and to understand others fast, our overall influence and persuasion power, and almost anything else you can come up with concerning leadership can be traced back into the brain. Your brain makes you a great leader. Do you allow it to do so?

Adaptive

One word can be used in order to characterize modern societies and collective entities within them: complexity. Societies, as well as organizations, as systems of infinite human interaction, can be seen as inherently complex. We usually ask students in our lectures to define complexity in the easiest and most convenient way they can. After discussion and various definitions suggested by students, we often conclude that complexity reflects a situation – any situation – that cannot be fully understood. In other words, complex is something that we cannot easily understand and, ultimately, control. It is important to note here that there is a critical distinction between *complicated* and *complex*. Complicated is something that depends on many factors that require, often, expert knowledge in order to explain it and place it under control. Complex is something that depends on many factors. However, these factors are almost impossible to predict, identify and to ultimately place under human control.

Complexity science emerged in the last 50 years and has changed the way we understand physical phenomena. Complexity and chaos theory influenced social sciences too, including management and leadership, mainly during the last three decades. In this context, complexity theory considers societies and organizations as complex adaptive systems comprised of numerous autonomous agents, which engage in a non-linear, unpredictable and emergent behaviour (McMillan, 2006; Psychogios and Garev, 2012). These systems have an innate capability to self-organize, since relationships within them are guided by continuous feedback loops (Kaufmann, 1993; Stacey, 2004). Similarly, chaos theory implies that social systems constantly evolve over time as they are sensitive to small changes that can make the whole system fluctuate with unpredictability (Wheatley, 1999; Cilliers, 2000).

Complexity in social sciences has become a strong paradigm. The realization that we live in a volatile, uncertain, complex and ambiguous (VUCA) world (Horney *et al*, 2010) is widespread. In order to better understand the VUCA world we live in, just think of one day in your life. Consider for a moment how much information you receive daily, related directly or indirectly to your job. Just think about your daily communication flux. How many emails you send and receive, independently of whether they are all relevant and, even more importantly, useful. How many times you interact with others on a daily basis, either through face-to-face communication or through phones and digital meetings platforms such as Skype. How much time you spend reading job-related news, visiting relevant websites etc and, furthermore, how long you spend each day on social media that may be of professional importance. Finally, consider the total information you receive from the wider media landscape. Now, think that the same is happening to most individuals, in one way or another, on this planet. The resulting web of interactions is mind-blowing. This is just a small thought exercise for the world we currently live in.

The world consists of dynamism, fast-paced evolution, unpredictability, huge uncertainty and, of course, endless information. Without any doubt, our VUCA world is an *infocratic* one. If the previous paradigm of social organization was *bureaucracy* the new paradigm is *infocracy* (Clawson, 2011). This paradigm claims that the source of power is no longer hidden within positions and offices. Because of the information explosion we experience, it is redistributed all over our organized societal entities. Infocracy requires a new approach, oriented towards continuous adaptation and not towards a single optimum outcome as in the bureaucratic paradigm. Infocracy claims that everything happening around us is somehow being co-created by all of us, even if some have more influence over the result (Stacey, 2010). In other words, the new paradigm elevates people as the critical component of any social system. People, and especially leaders, need to understand that they play a critical role in formulating the VUCA world around us. They are not passive actors. To do so though, they urgently need to develop new skills of adaptation.

Brain science brings a message of hope here. This hope is called *neuroplasticity*. Our brain does not go unchanged during our life. On the contrary, it changes constantly, every single day. Neuroplasticity is the proven ability of the brain to change, be trained, adapt, grow new neural connections or degrade the existing ones. All these depend, apart from the apparent genetic factor, on ourselves and the environment we live in and how we interact with it. Even in our adult and later years of our lives, our brain constantly moves (Hood, 2014). Consequently, the brain is the key to the new kind of leadership required to continuously adapt in a complex world.

Leadership

Leadership is an extremely popular term with high impact on our societies and organizations. Indicative of this fact is that searching for the terms 'leadership' and 'leadership in Business' in Google returns almost half a billion results for each. When you type 'leadership in business' you receive approximately 1.5 billion results. Leadership matters. Since the beginning of humankind and the development of primitive communities into the first civilizations, leaders played a significant role. But why do people believe so much in leadership? A critical answer can be found in psychology. According to Maccoby (2004), people are keen to follow a leader both for rational and irrational reasons. The former is about dealing with the unknown, which usually means people's future. Leaders can infuse stability and certainty when dealing with uncertain situations. They can give hope and set a clear path in front of us. However, irrational reasons seem to be more critical. People follow a leader, often without being aware of why they do so, because it is a natural response to difficult situations. This automatic willingness to follow leaders emerges from emotions and images in our unconscious mind that are projected onto our relationships with leaders (Maccoby, 2004). Leadership is an evolved state in our brains.

There are thousands of studies attempting to understand, explain and, paradoxically, somehow 'control' the phenomenon of leadership. Those exploring the literature on leadership (Pierce and Newstrom, 2014; Obolensky, 2014;

Western, 2013; Stacey, 2012; Psychogios, 2007) categorize it in four main approaches or leadership styles:

1. The autocratic-controlling approach: where the leader seeks and controls resources to enhance efficiency.
2. The motivational-engagement approach: where the leader focuses on relationships and on motivating others in order to enhance results.
3. The transformational approach: where the leader aims to transform the culture of an organization, attempting to achieve better outcomes.
4. The adaptive approach: where the leader sees organizations as complex systems that are evolving and therefore trying to constantly adapt.

There is an evolutionary pattern above. Leadership started with control, moved to motivation, then to transformation and finally to adaptation. Our approach, as explained in the book, does not simply belong in the fourth style. It strives to expand it. Adaptive leadership deals heavily with systems, and places them in the centre of analysis (Obolensky, 2014). It is the systems that are adaptive and leaders have to operate in a constantly changing environment within those systems. In our approach, the focus is not on systems or the external world, but on the brain, the internal world. For us, it is the brain that is adaptive too. And it is this phenomenal capacity of the brain to change that brings leadership to the forefront of creating better organizations and a better future for all.

We argue that modern leadership requires practical judgement, or what the ancient Greek philosopher Aristotle called *phronesis*. Phronesis is linked to judgement resulting from context-dependent actions, as these are embedded in our daily local experiences (Stacey, 2012). *Phronesis* is a way of thinking, knowing, acting and living our daily lives (Antonacopoulou, 2012). It is acquired through practical experience and it can be understood best when we take a reflective stance on this experience (Stacey, 2012). In turn, it helps us to formulate judgements and influence the way we behave (Antonacopoulou and Psychogios, 2015). We strongly believe that by understanding how the brain works and by accepting our ability to alter how the brain reacts to stimuli, we are in a better position to reflect on experience and formulate judgements, and therefore, to adapt our behaviour better towards ourselves and others. The BAL approach is critical in doing so.

The BAL approach in brief

The BAL approach consists of four pillars. These are the four main groups of ideas, scientific insights and practical recommendations that we have gathered, organized and used as a comprehensive brain-based approach in our business, managerial, educational and personal lives. In creating this approach, we used a large number of scientific and business literature focusing on the latest possible insights and combining them with classic ones. We have used academic studies, examples, case studies, personal stories, professional opinions and practical tips from around the world. But, most importantly, from our own consulting, research, training, coaching, entrepreneurial and managerial endeavours. We briefly describe each one.

Pillar 1: Thinking

This first pillar reflects the cognitive function of our brain. It is the longest pillar in this book because leaders and businesses are in great need of learning how to think in a brain-based way. If they do so, they can release the true power of their analytical thinking and achieve more. Becoming a brain-based leader first and foremost means maintaining and increasing willpower. The why is plain and simple: leadership is about keeping your brain strong and able to deal efficiently and effectively with demanding situations. Your thoughts and decisions will never be clear, relevant, useful and inspiring if you waste your brain energy. Learn how to save and use your willpower, avoiding the pitfalls of ego depletion, burnout and multitasking. Awareness of cognitive biases is of equal importance. Economists, psychologists, business authors and consultants alike see biases as the main evil of non-rational thinking. A strong brain, with plenty of willpower, is no good to anyone if it is also full of illusions. So, leaders need to be constantly aware of biases in order to consciously decide when these biases are helpful (sometimes they can be) and when they are not. Beware of biases and use them to your advantage when needed. Similarly, beware of automatic pattern recognition that can be misleading, and ask great questions to reveal the key elements in each situation – questioning is a core leadership skill. Adopting a deep and meaningful personal purpose helps leaders to keep their thoughts on track. At the same time, the ability to find and get into the optimum mental state of flow will help leaders boost clarity, effectiveness and productivity of thinking. Creative thinking emerges as the most important type of thinking since resourceful problem solving is necessary when dealing with disruption and constant change. Memory often appears in surveys as the first cognitive capability that people want to improve. A good working and long-term memory always impresses people around us and often helps win arguments. Leaders need to learn how to improve and use their memory. Along with memory, they also need to adopt an open, positive, growth-focused mindset that will keep them sharp and on top of any unexpected situations.

Pillar 2: **Emotions**

The second pillar reflects the emotional life of our brains. Humans are not always fit for understanding and articulating their own emotions and this creates a huge challenge for leaders, taking into account that what moves us are emotions, not thoughts. We provide concrete models on naming and categorizing core emotions and how we can recognize them in ourselves and in others. Being able to spot the right emotion early and to deal with it accordingly is essential for better leadership. Emotional equations and emotional agility can perform miracles for leaders, when they are used appropriately. We explore emotional styles, the permanent ways in which we emotionally react to stimuli, and how to identify them. Most importantly, we focus on how to change emotions to fit a modern leadership style. Leaders also need to examine their moods, or semi-stable emotional states, because a wrong mood can destroy the collaborative spirit of a team, a department or a whole organization. Find your positive mood and transmit it to those around you: you will reap immediate results.

Pillar 3: **Brain automations**

The third pillar reflects the automated responses and protocols of our brains. As our decisions and actions are primarily based on brain processes beyond our conscious control, rooted in deeper brain structures, it is imperative that modern leaders understand and use these processes to their advantage. Priming, the process by which leaders can nudge their brain and the brains of others towards a specific decision or behaviour, should be applied daily. Through priming, leaders can improve productivity, creativity and social connectivity. Similarly, habits, the most efficient way for the brain to save precious energy, should become strategic elements for enhancing leadership capacity. As patterns of repeated behaviour, habits in working environments are representative of the culture of the organization and need to be carefully approached. Recent literature clearly suggests that changing habits and creating a set of favourable ones, both for ourselves and others, is not difficult if we follow the right process. Destroy negative habits and create positive ones to boost your leadership effect. Furthermore, our brains have evolved over millions of years to deal with the challenges and opportunities of the physical world around it. Thus, it has developed unique ways to automatically interact and respond to the changes in its surroundings. In this respect, we suggest that the physical environment or physical context around us, subconsciously but significantly, affects leadership decision making and behaviour. The modern leader needs to understand the interaction between the brain and the physical world and use physical spaces to obtain desired behavioural responses.

Pillar 4: Relations

This fourth and last pillar reflects the social aspect of our leadership lives. No leader can inspire action if others do not connect with them in deep and meaningful ways. In order to connect, leaders have to fully appreciate the socially based origins of consciousness and adopt a 'collaborate first and then reciprocate' attitude in business interactions. Do's and don'ts are provided and modern leaders are encouraged to develop their own radar for identifying and connecting to people around them. This skill alone will dramatically increase your leadership effectiveness. Utilizing your brain's tendency for imitation in order to create and maintain an effective corporate culture, and establishing and managing extensive human networks, both through strong and weak ties, will elevate your leadership status. Furthermore, managing the art of clicking with other people by demonstrating presence, warmth and strength, will help you create powerful connections with important individuals inside and outside your organization. However, if you do not make sure that the relevant brain chemicals are in a collaborative mode, you will not be able to adapt your brain for networked leadership. Relations are built and nurtured through communication. Communication for us is about creating a desired behavioural response. Thus, persuasion is a core leadership skill because modern leaders need to influence behaviours to achieve their organization's goals. In this direction, we need to 'speak' to all three main brain functions, think, feel and do, if we are to induce and meaningful behavioural change. So we need to direct rationality, motivate by strong emotions, and tweak the environment for desired habits. To get maximum results both for individual and collective brains, leaders need to use, when possible, the principles of persuasion; specific words or phrases that can fast-track their influence to others (such as 'because'), compassionate conversation and the six key stimuli of getting a brain's attention. Your increased persuasion will impact positively both on your and other people's work.

Now, get ready to immerse yourself, and your brain, in the BAL approach.

Keep in mind while reading the book

1 Although neuroscience is the leading source for the BAL model we have also used psychology, sociology, behavioural economics, management and leadership science, and even marketing and communications, in order to explain and support our approach. Today, modern leaders need all the help they can get, wherever this comes from.

2 Use the references included to expand your knowledge on the issues you find most interesting. It is impossible to expand on all the topics, concepts, examples and recommendations in a single book. However, we have provided an extensive and comprehensive list of sources that we urge you to explore.

3 The short stories at the beginning of all chapters are real. The stories depict leaders in difficult business situations. All these stories come from our personal work with leaders and managers throughout the world. We have changed the details so that they cannot be recognized, but the specific challenges remain intact.

4 All exercises and recommendations in the book have been tried and tested over the last 10 years we have spent building the BAL approach. Experiment with as many as you can at appropriate moments and find the ones that work best for you and your team. Go for quantity as well as quality.

5 This is a book for practitioners by practitioners – not a book by neuroscientists for leaders. This is an important distinction. We do not provide details about the brain for its own sake. We have handpicked and carefully synthesized information we found to matter the most in our managerial, consulting, teaching and research practices. Contrary to most books on the subject, we have not used a single brain image in this book. Leaders have a huge need to be brain-friendly, but they do not need highly detailed accounts of brain anatomy. Although we do provide specific information on crucial brain functions, we always focus on practicality. This is what the BAL approach is all about.

6 The brain is still full of surprises so keep an open mind. Neuroscience, and other fields that deal with human nature, and especially with human behaviour, are still evolving. Technology and scientific inquiry will push the boundaries of understanding, and more ground-breaking insights are about to emerge. So, never stop searching, never stop asking, and never stop learning.

7 Don't just read, do. You learn more, and learn more quickly, through actually doing rather than just reading. Start applying the ideas and recommendations as soon as you feel like it. Experiment first in low-risk situations and slowly but surely move to more important tasks. We do not expect you to utilize them all going forward but we do expect you to enjoy exploring most of them.

8 Forget *homo economicus*. This out-dated view of humans, supported for centuries by some economists, is over and done with. Humans are not rational and they never will be. But this is actually a great thing. Rationality plays a crucial role in decision making, but without the moral compass of emotions and the steady guide of empathy we would just be cold psychopaths. Great leaders are not that. They are caring, understanding, passionate and visionary beings (with all of those characteristics or a combination of them) that move them forward no matter what. If rational, they would have stopped many times along the way.

9 It is about you personally and the people around you. The brain adaptive leader changes not just one brain, but many. Leaders who change their brains inevitably change the brains of those close to them, becoming an unstoppable force of progress. No leader is an island. Leaders are powerful social actors with a huge responsibility for improving everything and everyone, all the time.

10 We are all leaders. Front-line employees, middle managers and supervisors, unit and departmental directors, chief officers and CEOs: all leaders. The era of non-leader professionals is over. Complexity, disruption, abundant information and the most demanding clients ever have made the old model obsolete. Whatever the position, lead as a true brain-based leader and you will bring about change that matters. A lot of it.

It is time to move on to the first pillar.

References

Antonacopoulou, EP (2012) Leader-ship: Making waves, in *New Insights into Leadership: An international perspective*, ed H Owen, Kogan Page, London

Antonacopoulou, E and Psychogios, A (2015) Practising changing change: How middle managers take a stance towards lived experiences of change, *Annual Meeting of the Academy of Management*, August 2015, Vancouver, Canada (Conference Proceedings – Paper 14448)

Cilliers, P (2000) Rules and complex systems, *Emergence: Complexity and Organization*, 2 (3), pp 40–50

Clawson, J (2011) *Level Three Leadership*, 5th edn, Prentice Hall, London

Gorgiev, A and Dimitriadis, N (2015) Upgrading marketing research: Neuromarketing tools for understanding consumers, in *Trends and Innovations in Marketing Information Systems*, ed T Tsiakis, Business Science Reference, Hershey,

Henson, C and Rossou, P (2013) *Brain Wise Leadership: Practical neuroscience to survive and thrive at work*, Learning Quest, Sydney

Higgs, M and Rowland, D (2011) What does it take to implement change successfully? A study of the behaviours of successful change leaders, *Journal of Applied Behavioural Science*, 47 (3), pp 309–335

Hood, B (2014) *The Domesticated Brain*, Pelican Books, London

Horney, N, Passmore, B and O'Shea, T (2010) Leadership agility: A business imperative for a VUCA world, *People and Strategy*, 33 (4), pp 34–38

Kaufmann, SA (1993) *Origins of Order: Self organization and selection in evolution*, Oxford University Press, Oxford

Maccoby, M (2004) Why people follow the leader: The power of transference, *Harvard Business Review*, Organizational Culture Magazine Article, September 2004. Accessed at: https://hbr.org/2004/09/why-people-follow-the-leader-the-power-of-transference

McMillan, E (2006) *Complexity, Organizations and Change: An essential introduction*, Routledge, London

Obolensky, N (2014) *Complex Adaptive Leadership: Embracing Paradox and Uncertainty*, 2nd edn, Gower Publishing Limited, Farnham

Pierce, TJ and Newstrom, WJ (2014) *Leaders and the Leadership Process: Readings, self assessments and applications*, 6th edn, McGraw-Hill, New York

Psychogios, GA (2007) Towards the transformational leader: Addressing women's leadership style in modern business management, *Journal of Business and Society*, 20 (1 and 2), pp 169–180

Psychogios, A and Garev, S (2012) Understanding complexity leadership behaviour in SMEs: Lessons from a turbulent business environment, *Emergence: Complexity and Organization*, 14 (3), pp 1–22

Rock, D and Ringleb, Al H (2009) Defining NeuroLeadership as a field, *NeuroLeadership Journal*, 2, pp 1–7

Rock, D and Tang, Y (2009) Neuroscience of engagement, *NeuroLeadership Journal*, 2, pp 15–22

Rosenberg, MJ and Hovland, CI (1960) Cognitive, affective, and behavioral components of attitudes, in *Attitude Organization and Change: An analysis of consistency among attitude components*, eds MJ Rosenberg, CI Hovland, WJ McGuire, RP Abelson, and JW Brehm, pp 1–14, Yale University Press, New Haven

Stacey, RD (2004) *Strategic Management and Organizational Dynamics: The challenge of complexity*, 4th edn, Prentice Hall, London

Stacey, RD (2010) *Complexity and Organizational Reality: Uncertainty and the need to rethink management after the collapse of investment capitalism*, 2nd edn, Routledge, London

Stacey, RD (2012) *Tools and Techniques of Leadership and Management: Meeting the challenge of complexity*, 1st edn, Routledge, London

Western, S (2013) *Leadership: A critical text*, 2nd edn, Sage, London

Wheatley, JM (1999) *Leadership and the New Science*, 2nd edn, Better-Koehler Publishers, San Francisco

PILLAR 1
Thinking

Powerful brain, powerful leader   01

Friday's business plan meeting

It is Friday afternoon and everyone has showed up for the meeting. The whole week has been extremely demanding as the team have been finalizing the business plan for the next year. Working until late at night has been the norm during this week. The last important details on key strategic issues are to be clarified in this meeting and the tension in the room is evident. Everybody is physically and mentally exhausted and can't wait to get it over with. Although business planning has always been demanding, this year the pressure has been higher than ever due to new market challenges. This business plan could make a big difference for the company by creating real growth or it could perform below expectations, seriously damaging both the company's and the team's reputation. The team leader enters the room. She has always been a believer in leading by example so this week she was the first to enter the office early in the morning and the last to leave late at night. Her deep-rooted belief is that if they all work harder and longer they will succeed. She takes a look at the agenda then at the strained faces around the table. The meeting begins.

Implementation arguments quickly evolve into confrontational lock-ins. Disagreements on strategy turn into nasty personal comments. And the widespread dissatisfaction progresses by the end of the meeting into a few emotional breakdowns. Although the team ultimately manage to deliver a business plan that achieves its main goals, the team spirit that had been so carefully and expensively built over the last year has been irreversibly damaged. Relationships never recover and two key people leave the company within the next few months. Surprising as it may sound, the overall outcome of that Friday meeting had already been decided before it even started. Not by the people in that meeting. Not even by their leader. But by their brains.

How many times do leaders and managers face situations like that? Formal meetings, important presentations, feedback sessions or simple decision making are frequently performed by strained and overworked people. Regardless of the cultural stereotype of the all-powerful leader who is constantly on top form and eternally in total control, science shows that brain strength has limits. If those limits are ignored, self-control can deteriorate fast, with dire consequences for all involved. Willpower is a key leadership characteristic and one that has to be carefully nurtured to produce desirable support for both clear thinking and better management of emotions. And it all starts in the brain.

The energy-devouring brain

The brain consumes more energy than any other organ in our bodies. According to a *Scientific America* article in 2008, the brain takes, in total, more than 20 per cent of the whole energy available to the body (Swaminathan, 2008). Although it represents only 2 per cent of our body mass it consumes one fifth of the oxygen and one quarter of the glucose (Foer, 2012). If it were an organization this could mean that the brain represents a department with just 2 per cent of the total workforce, but one that demands 20 per cent of the company's financial resources. Naturally, every decision in this department would be taken with maximum efficiency and effectiveness in mind. Prioritization would be the name of the game and every action would be painstakingly weighed by its impact on the survival of the whole organization. The brain behaves in exactly the same manner. It prioritizes body functions that ensure its survival, redirecting energy to those strategic areas from more exotic and luxurious functions such as analysis and forecasting, if and when needed. Going back to the organizational analogy, it is not very different from the fact that many companies, when coping with diminishing sales, focus on and commit more resources to those areas believed to be essential for the survival of the business. At the same time they take resources away from more elaborate and risky future projects and from expensive advertising campaigns.

Going a step further, neuroscience reveals that the brain consumes most of its energy on automatic maintenance systems rather than on executive, higher cognition functions. The brain consumes almost 90 per cent of its energy in a calm state, when people are not asked to do much thinking. So, it actually has a small portion of its energy available to devote to complex and cognitively demanding tasks such as managing a year-end meeting, debating the last details of a business plan and navigating diplomatically through the team's conflicting opinions. We go through our lives being confident that most of the brain's processing happens in the area we are aware of, our consciousness. This could not be further from the truth. Most of the power is used elsewhere, deep into our brains. The part that we believe offers us our main competences and skills in developing ourselves and our careers, our analytical problem-solving part, receives just a fraction of the total energy. This means that we have to be extremely careful and strategic about managing this energy. Otherwise we will not have much to use in our demanding, challenging and exhausting daily management tasks.

The main question leaders need to answer then is how much power they will allow their brain to channel into its problem-solving section when they need to perform their best in such conditions. Are they going to create a situation where the brain will be starving for power and automatically will be reallocating it to more crucial and deeper structures for its survival? Or are they going to manage their working and living environments in such a way that will allow them to perform within maximum brain-power capacity when facing adversity? The first option represents the fast lane to bad decisions, angry responses and failed meetings. The second option represents the safe way for leaders to enhance their decision-making abilities and control over emotional reactions in an effective and meaningful way.

The willpower muscle for leaders

Willpower and self-control are not infinite. Walter Mischel, in his book *The Marshmallow Test*, analytically describes and explains the famous Stanford marshmallow experiment, conducted by him in the late '60s and early '70s. Walter has demonstrated graphically that when we are exposed to situations where we need to portray self-control we consume brain energy fast and thus perform much worse in the next task that needs willpower. Children that demonstrated higher willpower when asked to restrain themselves from eating a marshmallow put in front of them, performed worse in the next test on self-control that followed immediately after the marshmallow one. On the contrary, kids that showed less self-control and ate the marshmallow fast in the first test, performed much better in the second one. How can this happen? Isn't willpower something we either have or do not have? Results of the marshmallow experiment look counterintuitive since we tend to believe that willpower is a key personality trait. Societies seem to separate people into strong and weak, whatever the situation, and people tend to distinguish leaders stereotypically as strong and weak based on specific prototypes that they have in their minds.

When we present the marshmallow experiment in the class and we ask our audiences, students or executives, about the outcome of the *second* test, almost unanimously they respond in a predictable way: the kids that showed higher willpower in the first test did better in the second test as well. When we reveal the actual results it usually takes them a few minutes to adjust to this new view on willpower. This is further proof that it is not easy to overcome long-held beliefs in human nature and behaviour even when scientific proof is so strong. When they do come around to the idea that willpower is like a muscle that you can strain, they are able to explain past experiences better and see the future in a more confident and clearer way. They tend to recognize specific cases in their professional lives during which they passed through an exhausting process of the brain that led them to think and act in a less effective way at a later stage.

> **Action box**
>
> Sit down, relax and then mentally go back to the last time that you experienced a critical and demanding situation at work that needed effort and fast decisions. Think about the things that consumed your brain energy and then think about what happened afterwards. Consolidate your conclusions based on the arguments above and think what you need to be aware of next time. Do it. You will be amazed by the results.

The brain will utilize the power it has to help the person perform the desired task appropriately. This task could be a negotiation, a pitch presentation or a one-to-one coaching session. But being an energy-saving organ, when it is asked to consume a lot of energy and the overall energy levels fall it will reprioritize and make sure that energy flows primarily where it is needed the most. And this is always its survival and body maintenance centres. Thus, leaders need to be always aware of their brain's ability to distribute enough power to its executive part when it is crucial for them to perform at their best. Otherwise, primal emotions, confusing thoughts and an inability to focus on the right things will take over and damage processes, relations and outcomes. *Ego depletion* settles in when your ego, as in your ability to direct your own actions, is so tired and depleted of power that it cannot function properly anymore. In essence, it is not in control any more. You are not in control anymore. Primal brain functions have completely taken over your body. No leader wants to be in a situation like that. Ever!

The good news is that the willpower muscle in leaders' brains can be improved in the long term. The leading researcher of ego depletion, Roy

Baumeister (2011), in his book *Willpower: Rediscovering the greatest human strength*, has long argued the benefits of strengthening our willpower muscle and the perils of not doing so. From all available recommendations from the science of self-control, our experience has shown that higher-level thinking, strong values and immediate feedback are three key strategies in companies for strengthening the leadership willpower muscle. Let us explain.

Higher-level thinking

Higher-level thinking is described as engaging in more abstract and creative thoughts or simply considering the big picture of a situation. This way of thinking, which is more conceptual, is related to 'why' questions and is the exact opposite of low-level thinking that asks more 'what' and 'how' questions. Experiments have found that higher thinking helps the brain perform better in willpower tests than low-level thinking. Although this scientific finding has profound practical effects for management and leadership it is largely ignored in companies. Modern organizations seem to be obsessed with implementation and quick wins. 'Doing' is preferred over 'thinking' since fast market shifts require fast company responses. This leaves little space for engaging in true conceptual thinking and abstract contemplation. Even when such thinking is exercised it is usually the privilege of people further up the hierarchy, if not reserved exclusively for the very top of the ladder. At the same time though, companies demand their front-line staff show ultimate resilience and fight the daily battles in the volatile marketplace with advanced self-control and exemplary willpower. However, this can only happen when staff at all levels are encouraged to discuss openly the big picture, are exposed to 'why' questions and are motivated to reflect on the causes of things. The goal is to help people synthesize the whole rather than focusing on parts of the whole. In order to do so we need to recognize the contribution of high-level thinking to retaining our brain power. By facilitating this process we can enhance decisions and actions that are based on their willpower and they will exercise more self-control, achieving objectives efficiently and effectively.

Strong values

Strong values have also been associated with increased willpower. Leaders wanting to always be focused and mentally strong need to develop meaningful and deeply-held values. Hiring for attitude and training for skill is such a popular strategy, because people with the right work ethic will behave desirably most of the time, regardless of what they are faced with or how tired they are. They will simply have more energy to devote to their preferred behaviour than those who do not have values or have questionable ones. Additionally, operating under strong personal values provides a blueprint to the brain of behavioural norms. So, in situations of low energy levels, when someone has worked very hard over the last few days and feels very tired, the brain will retreat to a habitual mode and replicate default behaviours to save power. These default behaviours, if guided by deep-rooted values, will align with aspirations and will decrease the possibility of undesirable reactions to external stimuli. Building strong organizational values that people really believe in can work in the same way and collectively improve the willpower in the company. But creating strong values goes beyond office branding and in-house mantras. It should run in the DNA of a company from the top to the bottom. Values have to become the absolute yardstick for measuring all behaviours and all performances. Only in that way will the brain learn that these are absolutely necessary for survival and will relegate them to its region of automatic behaviours. Thus, having the necessary energy for them in cases where energy is scarce. In our experience values become an integrated part of the corporation's collective mind, when following the ABC framework for infusing organizational values.

The ABC of creating organizational values

A is for authority. Organizational values are implemented in a top-down manner, from positions of high authority to positions of low authority and not the other way around. This is because the brains of people lower in the hierarchy will subconsciously copy and paste the actual behaviours of people seen as having more power than them.

B is for barometer. Organizations need to measure periodically the status and maintenance of values among their members. The two main variables to be measured are the *importance* of each value and the *performance* of each value. That is, we need to measure the potential impact of each value in our business (importance of the value) and how much each value drives behaviour in the company (performance of the value).

C is for consistency. Each value should not be used just for internal PR purposes and for creating a nice brand story. Every value should be highly integrated in all organizational processes and procedures in a formal and clear to all way. This can enhance their maintenance by creating appropriate habits.

Feedback

Feedback provides tangible checkpoints for the brain to adjust behaviour. Absence of self-control usually makes a person more compliant, more sensitive to impulses or even more agitated. Immediate feedback can help in restraining bad behaviour and even in re-evaluating the situation, leading to corrective action. For example, if you receive formal feedback that you need to work on your composure in order to be considered for a promotion, it can work as a strong incentive for checking your behaviour next time you are in a meeting. Similarly, you can ask a trusted member of the team to subtly signal to you when the meeting is getting hotter if you feel you are not always able to detect it yourself. On a broader scale, developing an organizational culture that is transparent and supportive, and with widespread processes for delivering fast and accurate feedback, can help employees' brains receive the right signals at the right time to adjust their behaviour and thus to show more self-control and willpower. The more important the signal the more the brain will give it the energy it needs to deal with it. So, are we sure we are receiving the right feedback at the right time? Moreover, are we giving this feedback the right attention and weight to assist our brain to behave accordingly? If not, then the power we will need to respond strongly and affirmatively to the signal simply might not be there.

The science of willpower provides significant insights on what it takes for a leader to have a strong brain, but it does not tell the whole story. In order to have a more complete picture of brain power and to make sure we supercharge our brain to deal with complexity, we need to avoid burnout syndrome.

Burnout syndrome

So you finally took your long-awaited holiday to that exotic destination! It was surely well-deserved since for the last year you were working non-stop with multidisciplinary and multinational teams on a crucial restructuring project. It was very exhausting for the entire duration of the project. You were often daydreaming of this vacation and the time you would spend away from everything, just enjoying the sun and the sea. You would forget everything for a while and you would return to the office as good as new, super-ready for the next task. The problem is that as soon as you got back in the office, regardless of the physical rest, the mental rejuvenation and the emotional recharge you experienced in your holiday, one week into the job you started feeling as tired as ever. Let's be honest. A lot of us have had this strange feeling and, even worse, it is very difficult to motivate ourselves to explain it, thinking that we get used to holidays and we need to find our old performative self. However, this is not how it should work. You got your time off, thinking of big ideas or of no ideas at all. What's going on? Where do the tiredness and all the negative feelings towards the job come from? Chances are they come from burnout syndrome, which decreases your ability to perform as well as you always wanted.

Feeling chronically tired, both physically and mentally, is not uncommon in modern organizations where new problems arise every day and the lessons learned yesterday do not always apply to the challenges of tomorrow. But burnout syndrome does not just depend on tiredness and exhaustion. Again counterintuitively, it also depends on core job conditions that if not taken care of can cause your brain to disengage and perform below everyone's expectations and mainly yours. Can you be a true leader when suffering from burnout? We don't think so either. How can a leader inspire and influence others when they are constantly exhausted, cynical or feeling ineffective or all of the above combined? Because these are the main symptoms of experiencing burnout. And how can a leader lead a team that is burned out too, experiencing these symptoms for months and months on end? It's almost impossible.

There have been times in our consulting and coaching careers that we have encountered companies with widespread feelings of burnout. Departmental directors, middle managers and front-line staff alike complaining constantly about being void of energy, having to do too much, too fast, with more tasks being added daily and without any mercy. Even worse, often they are not in a position to effectively describe the situation that they face. They do feel and understand what they are going through daily but words don't seem to be enough to communicate this. So they cry for help at any given opportunity. In such cases the corporate culture seemed to be like an energy vampire that depletes its employees of their mental powers on purpose. Digging below the surface, a leader can very easily uncover the true causes of burnout and take immediate actions to make sure that it no longer influences his performance. By doing so, leaders can confront and reduce any trace of cynicism, boost their self-confidence and feel more empowered to engage themselves and others in achieving their goals.

Physical exhaustion should be addressed in various ways. Holidays can be a good solution, but daily and weekly doses of relaxation can help much more. A leader has to be aware of the state of high adrenaline reached when working long hours for an extended period of time. Although it feels momentarily good, it does so to hide pains and inefficiencies of the body and mind. It makes you keep going when you probably need to take a break and re-focus. Creating a working environment that allows for breaks and relaxing moments is vital for dealing effectively with burnout. The impact of recovery or decompression sessions cannot be stressed enough. However, the deeper causes of burnout are harder to deal with. But they should be dealt with decisively if we want to calibrate our brains for leadership.

The key word in understanding and preventing burnout is *mismatch*. When mismatch is experienced for a period of time without being resolved, it can lead to one or more symptoms of burnout. This mismatch can appear in many forms:

- increased complexity of a problem and limited time to resolve it;
- difference between job description and actual skills;
- nature of task/organizational culture against personal values;
- high expectations to perform and lack of available resources;
- given responsibility over task but absolute lack of control over task;
- pressure for performance accompanied by lack of appreciation from the company;
- very close structural relationship with very bad quality of relationship (supervisor over supervisee for example);
- variance between organizational talk and walk (managers saying one thing and doing another);
- newcomers faced with highly unpredictable jobs; and
- perceived seriousness of work and impoliteness of co-workers.

The list unfortunately can go on forever and managers and employees around the world are welcome to add their own mismatches here. Even worse, many are faced with a combination of the above mismatches, which creates multiple symptoms of burnout to manifest simultaneously. All in all, a very toxic environment for leadership qualities to be sustained. So what do we do?

Action box

Think of a couple of complex cases that you have needed to deal with. Then try to find what the specific mismatch was that made you feel exhausted and less willing to continue performing. What are your conclusions? What can you do next time to avoid similar mismatches? Remember, the important point is for you to identify what to avoid doing next time.

Our brain is a social organ. The fact that the worst punishment in prison is isolation proves the point that our brain can deal with many things, but not with absence of interaction. Our brain grows and becomes better through social interaction. Without meaningful and constant interaction in the early stages of our lives we can end up with serious mental issues and deficiencies. Our brain develops and learns mainly through external stimuli and our social surrounding plays a very important role. Although the relational part of fine-tuning the leadership brain will be discussed in later chapters, forging the right social interactions at work is considered to be the universal medicine for burnout.

Through our extensive experience in internal communications we have identified a number of occasions where improved information exchange internally reduces the mismatches mentioned above. For example, strategically designed, two-way internal communications can help leaders listen to their teams and transmit the right information to the right people. Explaining comprehensively and discussing company values, objectives, processes and results is taken lightly in so many companies while it is the easiest way to deal with mismatches. Information exchange through open communication is necessary for any leader to fight burnout both individually and collectively. Even at a micro level, every good relationship within a company is based on timely, accurate and interactive communication between two people or within a team. This really is unavoidable. Furthermore, a smile can go a long way. Treating people with respect and spreading kindness around creates the right conditions for reducing the burnout effect in others and in return in you. Last but not least, prioritizing the right tasks in your day and in your team, and making sure everyone understands what is happening and why good measures are against decreasing mismatches. Decreased self-control and the burnout syndrome can damage your leadership ability for good. Knowing how to spot symptoms and how to treat them in a timely manner will help you drastically improve your leadership brain. Nevertheless, there is one more condition that threatens your brain strength and that needs to be addressed by the modern leader. Multitasking is probably the trickiest of the three because many managers and directors we meet in companies are convinced that it is actually a positive characteristic of leadership. Multitasking hides more dangers than we think.

Multitasking yourself into cognitive load

Complexity, in the form of multiple, unpredictable and dynamic forces shaping our everyday business reality, has increased drastically over recent decades. Technology and global competition are just two of the elements of the external environment that change in a fast and unforeseen way. If shifting demographics, global politics, emerging scientific insights, demanding customers and the fluid state of world economics are added it is not difficult to understand the unprecedented pressure that modern leaders experience. Results, new product pipelines, talent management, value to shareholders and quick wins are areas where they are expected to perform spectacularly without many times having the luxury of hiring more personnel. Consequently, many people in companies believe that multitasking allows them to deal successfully with this complexity by simply performing more tasks at the same time. Does it work?

It doesn't. Regardless of the popular belief that multitasking is an impressive skill of super-efficient people it actually decreases the brain's power, thus weakening leaders. This is because multitasking is related to higher cognitive depletion. The more tasks you put on yourself to do simultaneously, the more pressure you add to your working memory. Ultimately, the higher the *cognitive load* in your brain, the faster it gets depleted of its available power. This should not be a surprise to managers since the negative relation between performance and overload seems to exist in many other types of systems. An overflow of simultaneous input causes congestion: from electricity grids to roads and from concert halls to rivers. Why did we think that our brains are different?

It is true that adults have a higher capacity for managing different tasks at the same time than children and the elderly. It is also true that education and experience can generally enhance the capacity for multitasking. But up to a point. The information load we generate as a society today has increased at an unbelievable rate and much faster than the brain's ability to deal with it. As reported in the technology portal techcrunch.com, Eric Schmidt, the Google CEO, said in a Technology Conference in 2010 that the amount of information we collectively produce every two days is equal to all the information produced by humanity up to 2003, mainly boosted by the extreme speed by which user-generated data are posted online. This startling claim should raise the alarm for everyone in organizations, since our current approach to multitasking needs to be urgently revised. Instead of treating multitasking as the antidote to increased complexity, pressure and information we should consider it as one of its diseases. Multitasking is not helping our leadership brain make better decisions. It does not allow it to focus on the right issues and inspire people. Instead it creates an illusion of efficiency fuelled by an inner urge for feeling proud of how much is accomplished in a limited amount of time. Multitasking is really not about efficiency. It is about sensation and about masking shortcomings in core leadership characteristics. We also need to stop thinking of leadership practice as another task. It is NOT. A study by Sanbonmatsu *et al* published in 2013, titled 'Who multi-tasks and why?' provides a number of amazing insights that should challenge the positive attitude towards multitasking in companies. The study found that people who considered themselves great multitaskers actually had an inflated opinion of themselves. They performed very badly in multitasking tests. Apparently, they did even worse than people who did not consider themselves multitaskers and did not engage much in multitasking in their everyday lives. That is an amazing discovery. The more we multitask and the more we consider ourselves as fantastic multitaskers, the less effective we are in reality. If this is the case, then why do people do it? Often multitaskers are thrill-seekers hunting for the excitement that achieving many tasks could bring. People labelling themselves as serial multitaskers are more impulsive, embracing whenever possible the rush of doing many things at the same time. *Doing* does not mean *doing right* though. It might feel good and exhilarating but it does not make us better leaders. Leadership is everything else but not a task for a multitasker. Leadership is about communicating, is about showing the way, is about motivating, is about supporting and developing and these are not traditional job tasks. They are, or they should be, daily behaviours for managers. By doing various things in parallel and trying to achieve multiple targets simultaneously, we may end up with serious losses in the leadership game.

The study went on to reveal that self-proclaimed multitaskers demonstrated important flaws in controlling the executive part of their brain. The executive part is the one that controls our impulses, directs attention and focus, deals with our working memory, and manages our problem-solving and task completion functions. Cognitive load can inhibit all these, making leaders weaker and incapable of fulfilling their influential role. How can we trust and follow a leader who demonstrates low attention, confused working memory and limited problem solving? We cannot. This is why multitasking should be approached with extreme caution. Multitaskers are less able to manage distractions and direct enough focus on a single task at a time. However, with increased complexity and the unmanageable rate of information flow, distractions and noise are here to stay. Businesses need leaders who will not jump impulsively every time a new direction appears. Quite the contrary, they need people who can navigate with conviction in the sea of challenges and lead others towards the right direction. With the limited amount of energy in the brain available for the executive part it is extremely risky to weaken its important executive functions because of too many tasks processed at the same time. True leaders are those who use their brain power wisely and selectively to make sure that they have enough for focus, problem solving, impulse inhibition and reasoning. Otherwise, those functions will receive less power than needed to work properly, resulting in confused, disoriented and frustrating behaviours. Self-control suffers greatly.

Learning from real life: the multitasking manager

We specifically remember a case where we were advising a business development director of a large retail company in the South East European region. With global experience and many international successes under his belt he was confident about himself and his team. However, the shockwaves of the 2008 economic crisis strongly hit the region and he had to deal with many 'fires' at the same time. This was a new situation for him. Therefore he tried to take on more responsibilities, attempting to confront increased complexity with increased multitasking. He was working more hours with multiple tasks on his daily agenda. He became demanding towards himself and others and started creating more problems than the ones solved. It took him some time to understand that multitasking was part of the disease, not of the solution. This was because spreading his attention to multiple locations and hot issues simultaneously was bringing too much excitement and an overblown feeling of short-term accomplishment. It was obvious to us that the quality of his attention, prioritization and decision making was deteriorating fast. Fortunately, an early mishap and the analysis of the potential bigger consequences of his multitasking drive made him reconsider his approach. With our help, he:

- delegated more by empowering his closest people to take some of the crucial decisions as well;
- focused on one core issue per meeting by increasing the number of meetings but decreasing drastically their duration;
- organized his work in a way that allowed efficient grouping of related tasks;
- set time aside to reflect on and evaluate new information and new tasks before moving to execution;
- increased transparency of workflow across the team so everyone was aware of what was happening at any time;

- used feedback better from both inside and outside the company after every task was completed to make sure that things moved as planned;

- took corrective actions first before moving to a whole new task;

- avoided unnecessary distractions by minimizing exposure to non-crucial sources;

- created quiet sessions within the week for concentrating better on the task at hand; and

- adopted a new weekly schedule that allowed for the unexpected and reprioritization as much as possible.

The company weathered the crisis better than the competition and managed to sign a few more lucrative deals in the process. The company did well because the above ideas shifted the whole cultural environment inside the company, making other people follow similar routes since he was doing so. And the leadership game was won.

An interesting detail from this case is the fact that at the start of the high pressure period our client was actually doing very well with multitasking. The study of Sanbonmatsu and associates did show that non-multitaskers performed better in multitasking when asked to do it, than those proud of their multitasking abilities. This means that we will not necessarily fail in multitasking if there is no alternative to perform many tasks simultaneously for a short period of time. However, if this becomes a habit and we get hooked on the sensational high of multitasking then we endanger both ourselves and our companies. Having said that, there is a tiny part of the population that could be characterized as *supertasking* due to its ability to multitask without the associated cognitive shortcomings of multitasking. Neuroscience still cannot explain the phenomenon, so to consider ourselves as supertaskers without close scientific observation is not only unreliable, but dangerous.

Keep in mind

Multitasking, ego depletion and the long-term syndrome of burnout are serious threats to the modern leader and the practice of leadership. Brain power is crucial for leadership since it is directly related to increased self-control, meaningful engagement and effective task completion. Thus, a powerful brain creates a powerful leader and not the other way around. You need to embrace the fact that the guardian of your brain's power is you. So, choose to protect it and to direct its energy where it is needed the most.

> **Boost your brain: learn to distinguish tasks from leadership**
>
> Make a list of tasks that you want/need to accomplish almost daily in your job. Then make another list of your actions as leader. Then think what you are doing to accomplish the items on the first list and what you do in order to behave as leader. Then think which of these tasks are really critical and which leadership actions are equally critical. Then think how much time you dedicate to the task side and then how much time you dedicate to leadership actions. Do you leave your brain enough power in both groups? How can you improve your brain power reserves based on what you read in this chapter? Consider your conclusions and use them next time.

References

Baumeister, FR and Tierney, J (2011) *Willpower: Rediscovering the greatest human strength*, Penguin Group, London

Foer, J (2012) *Moonwalking with Einstein: The art and science of remembering everything*, Penguin Books, London

Mischel, M (2014) *Marshmallow Test*, Transworld Publishers, London

Sanbonmatsu, DM, Strayer, DL, Medeiros-Ward, N and Watson, JM (2013) Who multi-tasks and why? Multi-tasking ability, perceived multi-tasking ability, impulsivity, and sensation seeking, *PLoS ONE*, 8, N. 1.

Siegler, MG (2010) Eric Schmidt: Every 2 days we create as much information as we did up to 2003 http://techcrunch.com/2010/08/04/schmidt-data/ Posted: 4 August

Swaminathan, N (2008) Why does the brain need so much power? *Scientific American*, 29 April

Clear mind, strong direction

02

The pattern has been revealed to me

The moment of truth. The CEO is about to announce the yearly numbers in front of the whole team. It has been very difficult recently since the company is struggling to come up with competitive ideas. New players are eating into its market share like never before. The team is disappointed with the third quarter's results and is expecting better sales figures in the last quarter. This is because the CEO, an admired leader of the company for so many years, has taken a series of bold decisions to reverse the negative trend, decisions they all supported as his charm and conviction are as strong as ever. He has been very confident about improving results and has the right insights to prove it.

At the beginning of the year he engaged a research agency to help him understand the situation better. It was the first time in his long career that he could not read the market as effortlessly as he always had done previously and he wanted to go deeper. Together with the agency and a few trusted directors they designed an ambitious study that aimed to measure all critical aspects of the business, internally and externally. The study took months to design and implement but six months ago, when the numbers came in, it became very clear to him what needed to be done. Although his team struggled to put everything together and see the big picture when the complicated and multilayered results were presented, he immediately got it. He feels young again, rejuvenated by his ability to see through all the graphs and capture the meaning of it all. He sees the pattern clearly. Full of confidence, he dictates to his team what they have to do as of tomorrow. When the procurement director expressed her reservations, both he and the other directors calmed her down and reassured her that it would all to be OK. They could all see it by now anyhow. The pattern is finally clear and it feels so good. However, when the CEO entered the room to announce the results at the end of the year his facial expression already said everything: the negative numbers had not gone away. It took two years for the company to recover and the people to understand.

How many times do managers and leaders experience the invigorating feeling of pattern recognition? How often do colleagues shout 'a-ha' in a meeting, having just conceived the ultimate truth from the numbers on the screen? And how many times were those patterns as far removed as possible from reality? The brain loves patterns. It does so because it makes it feel secure in situations of high uncertainty. If there is danger out there and you are not sure of what to do you are not going to move much. Not moving is more dangerous than moving, so the brain will try to find clues to quickly figure out the best option and to go for it. And it will figure this out regardless of the quality of the clues. So, problems arise when the selected course of action was based on the wrong understanding of the situation. Although happy to move, we can end up much worse than before the action. Leaders should take decisions being aware of such brain traps in order to make sure they are not fooled by them. Only in that way, will they not be fooling others with their actions as well.

The evolution of the survival-obsessed brain

We go about our lives being conscious. Well, most of the time that is and when we are not asleep. Being conscious gives us the feeling of being in control since we can observe our thinking and see when and how we make decisions. It is not by accident that our species is called *Homo Sapiens Sapiens* or 'the wisest of the wise'. We are the only species with advanced self-awareness and language capabilities. Although other mammals are also self-aware, such as chimpanzees and dolphins and even the humble magpies, humans are unique in the animal kingdom in engaging in sophisticated and elaborative thinking. This uniquely superior capacity, not just to think but also to reflect on our own thinking, has long been hailed as the pinnacle of what makes us human. 'I think therefore I am' said Descartes in the 17th century while much earlier Aristotle talked about rationality being a distinct human ability and thus humans being rational animals. Even art glorifies our thinking processes, as best embodied by the famous bronze statue *The Thinker* by Auguste Rodin, which was created at the turn of the 20th century. Somehow, it has become an axiom that thinking is controlling the human brain, or has the ability to do so, since this is what makes humans human. Thinking, represented by the executive part of our brain, became the absolute benchmark for human existence, the absolute skill and the absolute driver. Neuroscience though says otherwise.

The absolute purpose of our brain is not thinking. As we usually like to say in our lectures, the ability to use thinking exclusively and every time, is in reality wishful thinking. Thinking is the means to an end and the end is survival. And if survival is the aim then the brain will use all its functions to support it in an integrative and definite way. The brain assesses a specific situation and based on the urgency and importance to survival it will engage different functions to fit its purpose. Thinking is rarely a priority for the brain when reacting to a situation. This is because the executive brain functions, the ones that control emotions, forecast the future, estimate consequences and do calculations came to be much later in evolution. Approximately 1.9 million years ago, when the homo species first appeared, the brain started developing more sophisticated frontal lobe areas related with advanced language skills. So far so good. However, brains did not first appear in humans. Brains as concentrated nerve cells controlling bodily systems and behaviour exist in nature from much earlier. In the anatomy of the human brain, the older structures, usually referred to as the original brain, and the middle structures, usually referred to as the limbic system, exert much more influence on our behaviour than the newest addition, the neocortex. Whenever we ask people in our seminars to draw the brain they actually draw the neocortex, which seems to be the image most of us associate brains with. The real influence of the brain though comes from deeper inside.

The older or original structures are also called reptilian for a good reason. They represent the more basic functions of our brain such as controlling balance, heart rate, breathing and temperature and are found in the brain stem and cerebellum – very similar to a reptile's brain. As explained by Dan Hill (2010) in his book *Emotionomics: Leveraging emotions for business success*, this type of brain has existed for approximately 500 million years in nature. The limbic system, or border system, between the brain stem and cerebral hemispheres, includes functions controlling learning, memory, motivation and emotion. This type of brain has existed for approximately 200 million years in nature. Last and probably least, the neocortex is the youngest of the three, acquiring its more advanced features less than 2 million years ago. How did we come to believe that the youngest is the strongest? Examining neural patterns in the brain challenges the notion that thinking is on top. There are 10 times more signals going from the limbic system to the neocortex than the other way around. Most surprisingly, the brain activity in the neocortex is just 5 per cent of the total brain activity. Thinking comes last by all measurements in the brain's priority for survival. Evolutionary speaking, this makes a lot of sense.

Whenever we present those facts to business executives they are astonished. So deeply rooted is the belief that our brain is a logical machine that any proof of the opposite creates surprise at best and more often disbelief. This realization has a profound effect on the way leaders make decisions in companies. When it becomes clear that our brains create desired responses to stimuli driven mainly by very old and hidden forces, leaders start changing their view on how they make decisions. Their confidence in naturally and automatically occurring logic is shaken. And then they start afresh.

Beware of patterns bearing gifts

So why do we like patterns so much? Employment and customer trends, the latest technological developments, macroeconomic cycles, changes in investment and political decision making, wherever we look there are patterns waiting to be discovered. Patterns that can make or break our business, reinforce or damage our leadership status. We just cannot help it because patterns make us happy. Dopamine, a very influential reward chemical in our brain, is to blame. Whenever we spot a pattern, our brain matches stimuli with memory. This means that whenever a new situation appears, such as the latest data on a spreadsheet or new competitors in our market, our brain will desperately search our memory to match this new information with existing previous information. Finding a complete match or creating a new pattern by combining various older information will have the same effect. The brain will be rewarded with a rush of dopamine. No pattern means no explanation. No explanation means more uncertainty. Uncertainty means no survival. In short, patterns create the feeling of control because they increase, realistically or unrealistically, our chances of survival. It is a very effective evolutionary mechanism that pushes our brains to always look for patterns, thus ensuring our continuation as a species.

Dopamine is not to be taken lightly. It is a very strong reward hormone and neurotransmitter in the brain that creates immense pleasure when released. Not all pattern recognition releases large amounts of dopamine though. Pattern recognition is a normal and automatic brain function that helps us recognize familiar faces, symbols and music. It would be catastrophic if every simple pattern recognition during the day was accompanied by significant amounts of dopamine. Immense pleasure actually accompanies pattern recognition when the pattern arrives in a very difficult and unpredictable situation. When it comes least expected and when no one else sees it. This is when we get the most pleasure from it. This happens because of the 'prediction error' function of dopamine. Dopamine is mostly released when rewards are least expected and surprising, not when it is business as usual. The example from the beginning of the chapter showcases this mechanism in a very clear manner. It was when the situation was critical, the stakes high and no one had a clue what to do that pattern recognition offered its most rewards to the CEO. Alarmed by his initial inability to read the signs and overwhelmed by the data in front of him, his memory retrieved old patterns to match the situation. When this happened he felt excited, rejuvenated and full of confidence.

Executives from around the world constantly explain to us the dynamic nature of their job. Unpredictability is on the daily agenda and complexity is involved in every major decision they have to make. It is their responsibility to see the whole picture and offer solutions that will inspire others. At the same time they need to be as accurate as possible. So pattern recognition is a key aspect of their role in very challenging environments and a dopamine rush is waiting in every corner. However, leaders with a series of wrong calls eventually lose both their followers and their job. So being aware of this brain trap is more crucial than ever for people in modern organizations.

We cannot avoid seeing patterns in data, people's behaviour and market trends. This is our brain's evolutionary role in ensuring our survival. It is possible to avoid the high of spotting a potential wrong pattern by making sure we examine all possible viewpoints, with equal attention to diverse opinions. Pattern recognition will emerge from deep inside our brains. Double-checking and cross-checking should be immediate reactions to pattern recognition coming from the top of our brains this time. Always remember that the gifts of pattern seeking can be great but they can also be the Trojan horse of your leadership future.

Together we stand... or do we?

'There is no "I" in team' the popular saying goes. The obvious meaning of this statement is that when in a team you have to prioritize the team over yourself. The hidden meaning though is that teamwork can extinguish individual voices for the collective good. But what if these isolated voices were right and the team was wrong? You might recall that in the example at the beginning of the chapter, one of the directors voiced her concerns over the CEO's pattern revelations. But she was immediately shot down by both the CEO himself and his pattern disciples. Finally she went along with it, converting to the team truth soon after. After the disastrous end of the year she was one of the first to leave the company for another job.

It is not rare in companies to find teams enchanted by an idea, following it as a panacea for all or most of their problems. From a new ERP system to an expensive rebranding exercise and from a new high-level recruit to an innovative product that will be the definite game-changer, teams often fall into the trap of groupthink. We have encountered groupthink many times in our training and consulting engagements. It usually manifests itself first as a proven solution, then as a strong team commitment before finally revealing its true face, a dogma that no one has the right to challenge. There are a lot of studies, mainly in social psychology, on this phenomenon pioneered by the US psychologist Irving Janis. In his 1982 book titled *Groupthink: Psychological studies of policy decisions and fiascoes*, he views groupthink as an in-group unwillingness and blindness to realistically see alternatives of a decision and/or action. Policy and especially decision-making groups can become so cohesive and influenced by past success that their leaders can easily gain uncritical support from team members for poor decisions. There are numerous laboratory studies that have attempted to validate Janis's arguments. Many of them have confirmed that teams with a moderate amount of cohesiveness and more pragmatic collective confidence produce better decisions than other teams with low or very high cohesiveness. In contrast, the latter groups made very poor decisions influenced by groupthink.

The neurological aspect of groupthink is that again we witness the interplay between the deeper brain structures and the executive part, especially the domination of the old over the new brain. Driven by an ever-present desire for survival the brain will push forward agendas that favour group homogeneity over challenging the status quo. More so in cases where the beloved idea comes from the charismatic and all-powerful team leader and everyone else seems to enthusiastically go along with it. Groupthink threatens the individual brain with 'you are either with us or against us', which implies dire consequences if the latter. The choice is ours but most probably it is has already been taken by our brain.

Action box

In order to avoid or protect your teams from groupthink you can do various things:

- Assign one or two members of the team to critically evaluate decisions all the time. These members can be changed according to the particular case.
- Use different teams to explore the same issue and avoid communication across the teams during this process.
- Avoid using only experts to participate in the decision-making process. Use people who are less experienced or invite outsiders to be involved sometimes. That way you can enhance the introduction of fresh and unexplored ideas.

Try the actions above. Also, you need to think of other ways through which you can challenge the status quo of a well-established team, without risking losing team cohesiveness.

Brain misfires

Together with pattern recognition and groupthink, there are a large number of automatic brain responses to situations that make managers make wrong decisions and engage in misguided actions. Those deeper brain structures, the old reptilian brain and the limbic system, having survival as their top priority for millennia, apply fast, unchecked thinking. Being very powerful, they drag our behaviour along with their wishes, giving little chance for the executive brain to interfere strategically. But how deep does the rabbit hole go?

Actually quite deep. Cognitive biases are now well-researched in multiple disciplines (Haselton *et al*, 2005). They occur most often when urgency is heightened, information is confusing and our social surrounding suggests conformity. Our brain's natural limit in information processing causes automated protocols to take over, leading to specific behaviours. Rationality is bypassed. It takes the backseat while the driver is a subjective, closed, impulsive and self-confirming person. This person cares only about fast responses based on what is already known and comfortable. Risk is unacceptable. Thus, spending energy in explorations and experimentations is seen by the brain as taking away vital energy from more reassuring repeated behaviours. All in all, not a great route for aspiring leaders to follow.

The threat of cognitive biases is ever-present and can be expressed in any managerial situation. When you label people and apply stereotypes, when you are not open to new and conflicting information, when you trust your memory and experience too much, and when you place yourself in the middle of the universe, cognitive biases are at work (Ariely, 2008; Kahneman, 2011). The brain does not direct enough energy to the prefrontal lobe for a more objective analysis of the situation. It prefers to resort to less energy-demanding neural patterns and see things as they should be and not as they really are. And as we have discovered through our engagement with hundreds of companies in more than 20 countries, our modern business environment cultivates corporate cultures that push for biases to surface more often than not.

Corporate cultures favouring the biased mind

A social psychology study on a biblical story reveals how companies build cultures that favour biases. In 1973, Darley and Batson published their now classic study on the parable of the Good Samaritan. The parable is about someone helping a person in great need without pursuing any selfish goals. The two scientists run an experiment in Princeton asking their students to come to a building to talk about the Good Samaritan and another unrelated topic. Then they were instructed to run to another building for an exam. Some of the students were asked to hurry to get to the other building while others were asked to hurry less. On the way to the other building all students passed by a person asking for immediate help and looking like he really needed it. Who helped more? Surprisingly, dealing with the Good Samaritan story prior to encountering the man had no effect on students' reaction. What had a real effect on their behaviour was urgency. Only 10 per cent of students that have been asked to be very fast in getting to the other building stopped to offer help. While more than 60 per cent of the students that were not in such a hurry stopped to help.

Many companies globally have corporate values telling a similar story to that of the Good Samaritan. Customer-centric, solutions-oriented, team-supporting, employee-promoting, community-focused are values we constantly encounter in corporate materials around the world. What we actually see inside companies though is that urgency cancels those values, prioritizing self-centred, short-term goals. A tight deadline, the next big project starting now, the next morning meeting you need to prepare yourself today for, the latest numbers that were not as good as expected, the new director that wants more and more, and other emergency calls create cultures that nurture automatic behaviours, frontal lobe bypassing and cognitive biases. Regardless of the nice narratives of good deeds in businesses, the pressure of 'now' decreases the brain's ability to direct energy to vital thinking structures that can ask the right questions. Instead of asking we are just doing. Most importantly we fail to stop and ask the man in need 'how can I help?' or 'how can I make the situation better?' as in the Princeton experiment. In companies today, questions such as these addressed to our teammates, subordinates, bosses, people in other departments, and of course to ourselves are crucial to keep biases and knee-jerk reactions at bay. But to do this, we need to start asking and stop rushing.

A study on how arousal changes behavioural intentions is also very helpful in supporting corporate cultures that are more considerate and reflective. Ariely and Loewenstein (2006) asked students to answer questions on moral behaviour, instructing them to answer the same questionnaire later when in a state of arousal. The answers were significantly different, showcasing graphically that well-thought intentions can go out of the window when specific brain sections get aroused. We know to do the right thing when calm but when aroused we behave differently. The study focused on sexual arousal but results are not confined in this brain activity exclusively. Arousal involves the activation of our automatic nervous system and leads to increased heart rate, higher blood pressure and an extreme sensitivity to external stimuli. In such cases, rational thinking loses control and power lies with more impulsive and pre-programmed reactions. Our corporate narrative might be promoting all the great values of understanding, collaboration and reflection but urgency and the aroused/stressed state of mind we constantly find ourselves in at work makes us more subservient to our biases than ever. The answer is to ask more questions.

Asking is winning

The most reliable managerial weapon against cognitive biases is actually the simplest one: asking questions. A lot of them and all the time. Asking the right questions, and even some wrong ones, creates the appropriate conditions for challenging automatic reactions and thus revealing any biases that might be surfacing. Questioning is about getting to a mental state of observation prior to action. Inserting this intermediary step of questioning minimizes the immediate behavioural effect of biases. Instead of the brain jumping directly from a stimulus to a premature conclusion or to a misguided behavioural reaction, a series of questions has valuable stopping power. That is, it has the power to quickly interrupt neural processes and impulsive activity, redirecting more brain energy on thinking and reflecting functions.

The power of asking questions in companies and other institutions in our societies is finally being noticed. According to Warren Berger (2014), a major advocate of informed enquiring, our brain is hungry for information from the time we are born, with 40,000 questions asked between the age of two and five and 300 questions asked daily at the age of four. However, the rate of questioning peaks at five years old and declines soon after. The reason for this is the anti-enquiring culture we have built at home, in schools and at jobs. Instead of encouraging questions as one of the most natural ways for our brain to learn and change behaviour, we have neutralized it. When you are faced with the attitude of 'it is what it is' and 'this is how we do things here' you instantly know you are in an environment that is hostile to informed enquiry. But asking questions is just the beginning. Asking the right questions is of equal importance too. Schoemaker and Krupp (2015), based on their extensive corporate research, have categorized key questions asked by winning leaders into six groups:

1 Questions that make you think outside the box and uncover the hidden meaning of wider market trends.

2 Questions that help you explore future scenarios and analyse major external uncertainties for each one.

3 Questions that help you become a contrarian who examines every problem from multiple and diverse angles.

4 Questions that help you identify the right patterns by deploying multiple lenses for discovering difficult-to-see connections.

5 Questions that help you create new options by evaluating numerous alternatives and their unintended consequences.

6 Questions that help you learn from failure fast and use this as a source for immediate improvements and further innovations.

For them, the starting point to long-term leadership success is not in answers but questions. The list above is a very effective remedy for battling cognitive biases and limiting the brain's obsession with the known.

Action box

By using the framework above, try to develop some questions for each of these six types. Try to think of specific situations in order to develop these questions. Get some of your people/team members to do the same. Can you spot the difference?

Socrates famously said 'I know one thing, that I know nothing'. His questions were so powerful that in the end they resulted in capital punishment. Societies with a slow pace of change, as the ancient ones were, do not like challenges to the status quo. Companies of the previous paradigm, operating in stable industries, did not really foster a culture of enquiry. Today things are very different. Examples of CEOs turning around companies by the power of questioning, supporting and learning are now common. Frank Blake, the CEO of Home Depot in the United States, who guided the turnaround of the company in the very challenging times after the 2008 crisis, did so by knowing very little about the company and the industry when he took over the CEO position (Reingold, 2014). His commitment to healthy and productive enquiry at all levels and his recommendation for learning walkabouts in stores for his key managerial personnel led to the right changes that increased customer satisfaction and brought higher share value. The power of asking in action brings more energy to the rational part of your brain and minimizes the impact of neural shortcuts. What kind of leader do you want to be?

Enquiring with style

Asking the right questions is still not enough though. We also need to enquire with the right attitude in order to make sure that the outcome will be beneficial for all. We have experienced cases where directors practised the art of questioning aggressively and in a non-productive way. Many times this type of questioning creates an extremely uncomfortable environment where questioning looks more like the interrogation than collaborative thinking. Below you can see our questioning style matrix.

TABLE 2.1 Questioning style matrix

		Engagement	
		Active	Passive
Attitude	Constructive	The star enquirer	The good listener
	Non-constructive	The evil interrogator	The grumpy loner

We have identified four types:

- *The star enquirer*. The leader is actively engaging in conversations, often as the initiator and facilitator. All discussions are friendly and often creatively heated and there is no wrong answer. The aim is mutual development.
- *The good listener*. The leader is always open to answering questions and providing feedback. Initiation of discussions though is mostly done by others. Listening, guiding and deciding are the main characteristics but in a reactive manner.
- *The evil interrogator*. The leader is approaching everyone with difficult to answer and even offensive questions. The main purpose is to make others feel inferior and incapable. This leader is feared, not respected.

- *The grumpy loner.* The leader is isolated by their own choice, giving every signal possible to others to keep away. Makes people feel uncomfortable when disturbed and looks like everything is under firm control that should not be questioned. Questioning is associated with weakness.

Moving towards the star enquirer quadrant is essential for a leadership-calibrated brain. It might not be easy to change but it is necessary in order to fight biases and other mental shortcuts.

Action box

By using the framework above, try to evaluate yourself. Think about each one of these types and try to classify yourself in one or the other category. Write down what to do in order to place yourself in one or another type. Think what you need to do in order to be able to place yourself into another type. Write down your thoughts again. It is important to write down your thoughts since this can help you in developing much more structured and concrete thinking around these issues.

Learning from real life: the tale of two managers

- Manager A asks the right questions, at the right time, of the right people.
- Manager B never asks in order not to appear weak.
- Manager A encourages independent thinking.
- Manager B allows you to express any opinion in meetings as far as it confirms his.
- Manager A always double-checks data... and then once more.
- Manager B takes one look at data and moves on to implementation because the pattern is always evident.
- Manager A has a 'to-the-best-of-my knowledge' attitude.
- Manager B has a 'my-knowledge-is-the-best' attitude.
- Manager A exposes herself to the ideas of outsiders.

Clear mind, strong direction

- Manager B has only disbelief of the irrelevant views of outsiders.
- Manager A forces the team to individually and collectively reflect often on decisions and projects.
- Manager B has no time for reflection, only time to start another project.
- Manager A favours research that involves multiple samples and a variety of data collection methods.
- Manager B has used the same research approach for years because it has been tested and worked.
- Manager A attends conferences, industrial meetings and training with a productive, open-minded attitude.
- Manager B always comes back from an event with the phrase 'I knew that already, nothing new here'.
- Manager A follows top thinkers in diverse fields on twitter, blogs and publications.
- Manager B does not follow anyone (at least not openly); others should follow him.
- Manager A allows herself regular breaks during the day and periodical longer escapes to re-set her mind.
- Manager B believes that breaks are for the weak.
- Manager A takes interest when she spots a bias, in her or in others, always being polite and gentle to herself about it or to others.
- Manager B never accepts that an error of thinking came from his bias because his thinking is based on pure rationality. Everyone else is full of biases but him.
- Manager A considers every opinion carefully – not necessarily to the same extent though – even when it comes from a chance conversation outside the office with people not related to her work.
- Manager B never talks business to anyone outside the office, except to boast about a success.
- Manager A tests and uses the latest apps for productivity, collaboration and idea generation. She loves experimenting!
- Manager B uses emails and spreadsheets. He hates experimenting since this shows that people don't know what they are doing.

> Unfortunately we have yet to meet a person who stands exactly in between those two extremes. Although it is not necessary for someone to possess all the characteristics of Manager A or B, we constantly find that most characteristics will be in one or the other category. Those two managerial types correspond with Philip Tetlock's (2005) classic categorization of political forecasters as foxes and hedgehogs. In his study *Expert Political Judgment: How good is it? How can we know?*, exploring what makes people, political outcomes better at predicting he recalled an ancient Greek poet who famously said that foxes know many things while hedgehogs know only one great thing. In Tetlock's widely accepted view (Silver, 2012), people's general attitudes critically affect their thinking and prediction capabilities. A fox is open-minded, accepts responsibility for mistakes, always learns and sees the world as a complex and difficult to predict system. The hedgehog is stubborn, always seeking to confirm old beliefs, does not listen to outsiders and is happy to explain everything according to existing theories. Foxes, on aggregate, are far better forecasters than hedgehogs. Manager A is a fox. Her brain is better calibrated to become a great leader in dynamic and chaotic times. Manager B is a relic, heading for extinction if he doesn't change fast.

When the ape takes control

One of the most significant lessons we learned early on in our quest to understand the brain's impact on leadership, and on human behaviour in general, was the existence of significant survival protocols. The brain has its own ancient protocols deeply rooted into its hidden neural networks that are activated in extreme cases to ensure our survival. And we learned it the hard way.

One of us had a car accident at the end of the '90s. Nobody was seriously injured in the incident but remarkably enough the exact moment of crash was not experienced consciously because of a fast blackout of perception. Gaining consciousness a few seconds after the impact was crucial for taking the necessary steps to escape with minimum injuries. The big question though was who decided to shut down consciousness at the exact moment of the impact? Surely this was not a rational decision since everything happened in milliseconds. Searching for an answer led to the realization that the brain applies its own rules when it decides that a situation is critical. External stimuli travel through our senses to the brain which, faster than thinking, decides how to treat them. In the case of the accident, the brain decided to blackout perception in order to avoid clogging the system with a shock. If in shock, then behaviour might not have been optimized for survival (Goleman, 1998). We can call such protocols *hijacking*. But is the *hijacking* of executive brain functions by more primitive ones happening only in extreme cases? No. It happens every day. Modern leaders need to be particularly aware of a brain *hijacking* mechanism threatening their clear thinking on a daily basis, which is called the amygdala hijacking.

The amygdala is part of the limbic system of the brain which, as described earlier, is the system that deals more with memory, learning, motivation and emotion. Within this system the amygdala plays the role of the emotional guardian of survival, deciding if and when to interfere. Interference means taking over. The amygdala is generally associated with emotional learning since it matches external events with internal memory to choose the best possible action. If the external event threatens to evoke a strong unpleasant memory then it decides to take control by stopping rational lobes from overthinking. In dangerous situations there is not much time for debating and deliberating. Consequently, the amygdala is also associated with stress, anxiety, fear and aggression (Goleman, 1996). For leaders, this means that there are daily situations at work that the amygdala can consider as threatening and hijack higher thinking with important consequences. Those consequences include erratic behaviour, mistrust towards others, low self-confidence, impaired decision making and performance, inability to read signals (human and numerical), self-centredness, short-term thinking, misunderstandings and miscommunication, knee-jerk reactions and revengeful attitudes. Hardly the desired characteristics of inspiring individuals.

We are not defenceless against amygdala hijackings. There is a lot that can be done. Professor Steve Peters (2012) from the University of Sheffield has developed a mind management programme by which he helps athletes and other high-achievers to deal effectively with such hijackings. According to his approach, our internal 'chimpanzee', who courtesy of evolution is five time stronger than our conscious will, influences our behaviour by attacking our mind with poisonous and fearful thoughts such as *I cannot do this* and *they are all against me*. Typical statements that give away the internal chimp's intentions are:

> Things will not go as I would like them to – external issues.
> I do not feel fit for the task in front of me – internal issues.

The way to deal with the ape that lives in our head and has the power to override our clear thinking is to first accept him and avoid fighting with him (he is stronger). We need to observe all his/her moves and impulses and then to nurture his/her better side by making him/her feel less insecure in the long term. Finally we can manage him/her through a system of distractions and rewards. A distraction can be something that will not allow the chimp to express himself/herself fully, like entering the presentation room quickly without much last-minute stressing about it. A reward is about giving the chimp something he likes when the work he was opposing has been done, partially or totally. This could be something like having your first coffee break only after you reviewed the difficult report in the morning.

A more physical approach in dealing with hijacks is advocated by Dr Alan Watkins in his book *Coherence: The secret science of brilliant leadership*. Dr Watkins (2013) suggests a relaxation method through breathing that can calm the irregular heart beating of a mind full of amygdala-produced fear. Taking the time to breathe regularly, smoothly and through the heart on a daily basis can have a significant impact on stress reactions. It can grant the mind the right space to develop constructive feelings and thoughts. Ignoring the physiology of hijacks is a strategic mistake that reduces the ability to deal with them successfully. We always recommend to our clients to take a short break from a situation when they feel overwhelmed by anger or fear. The popular advice of never sending an email when angry, but rather leaving it for the next day, has its origins in the interdependence of physiology and psychology. Take a walk, breathe calmly and regularly from your chest, and allow your system the necessary time to flush out chemicals produced by aggression and fear. Those simple acts can do more than you think for boosting your leadership performance.

The key issue of managing hijacks is to be aware and separate them from you. If you treat such negative thoughts with a genuine interest rather than with a fatalistic acceptance, half of the job is done. Your thoughts are not necessarily you. Although you cannot easily decide what comes into your mind you can decide what to do with those thoughts, especially when your body is in a calmer state. It should be your clear mind that leads your people, not your internal ape.

Keep in mind

Our mind can never be free from irrational influences pushed up by the older brain structures. However, by resisting the dopamine high of pattern recognition and the gravitational pull of cognitive biases, and by being aware of the internal ape that wants to get the best of us, we can achieve a clearer state of mind. Constantly asking the right questions and creating a culture that values debating will increase the strategic role of clear thinking in our daily leadership challenges. This in turn will allow us to give clearer and more meaningful direction to our team.

> **Boost your brain: you and your ape**
>
> Think about the ape side of your brain. Make a list of things that you have done when the ape was in control. What were those situations mainly about and can you categorize them? Then make another list with potential actions as a leader through which you need to control the ape inside you. Consider your conclusions and be aware of them next time one of those situations arises.

References

Ariely, D (2008) *Predictably Irrational: The hidden forces that shape our decisions*, HarperCollins, London

Ariely, D and Loewenstein, G (2006) The heat of the moment: The effect of sexual arousal on sexual decision making, *Journal of Behavioral Decision Making*, **19** (2), pp 87–98

Berger, W (2014) *A More Beautiful Question: The power of inquiry to spark breakthrough ideas*, Bloomsbury, New York

Darley, JM and Batson, CD (1973) 'From Jerusalem to Jericho': A study of situational and dispositional variables in helping behaviour, *Journal of Personality and Social Psychology*, **27** (1), pp 100–108

Goleman, D (1986) *Emotional Intelligence: Why it can matter more than IQ*, Bloomsbury, New York

Goleman, D (1998) *Vital Lies, Simple Truths: The psychology of self deception*, New Edition, Bloomsbury, New York

Haselton, MG, Nettle, D and Andrews, PW (2005) The evolution of cognitive bias, in *The Handbook of Evolutionary Psychology*, ed DM Buss, pp 724–746, John Wiley & Sons Inc, Hoboken, NJ

Hill, D (2010) *Emotionomics: Leveraging emotions for business success*, Kogan Page, London

Janis, IL (1982) *Groupthink: Psychological studies of policy decisions and fiascos*, Houghton Mifflin, Boston

Kahneman, D (2011) *Thinking, Fast and Slow*, Farrar, Straus and Giroux, New York

Peters, S (2012) *The Chimp Paradox: The mind management program to help you achieve success, confidence, and happiness*, Vermilion, London

Reingold, J (2014) How Home Depot CEO Frank Blake kept his legacy from being hacked, *Fortune*, 17 November

Schoemaker, PJ and Krupp, S (2015) The power of asking pivotal questions, *MIT Sloan Management Review*, 56 (2), 39–47

Silver, N (2012) *The Signal and the Noise: Why so many predictions fail – but some don't*, The Penguin Press, New York

Tetlock, PE (2005) *Expert Political Judgment: How good is it? How can we know?* Princeton University Press, Princeton

Watkins, A (2013) *Coherence: The secret science of brilliant leadership*, Kogan Page, London

Higher performance, more followers

03

It's all for the sales... or is it?

She gladly accepts the new position and the responsibility of running the global sales operations. The CEO has made it clear that he trusts her fully. He has also stressed the fact that she is to develop the new sales strategy that will bring them in par with other global players. The future of the company is in her hands and she is fine with that. Throughout her life, whenever she had to deliver results, she never felt scared. Her confidence was based on her ability to prioritize correctly, roll up her sleeves and tackle the problem head on. And it always worked. But would it work now, for the most important role in her career?

It starts well. She develops the main strategic direction herself, with key inputs from her peers and a couple of seasoned consultants. The strategy is very well thought of, with great analysis, and is already accepted with praise by the board. It is a bold plan, including innovative approaches. Times are challenging and require fresh thinking and deep changes to solve new problems. What is left is to present the plan to the regional sales directors in order to start implementing it globally as soon as possible. She has done big and important presentations before. Nevertheless, this is *the* most important presentation in her career. Naturally, she stays awake at night, perfecting every sentence and carefully designing every slide. She rehearses and then she rehearses again. Before entering the hall she runs through the numbers once more in her mind and makes sure she has everything in the right order. She is aware of the fact that such changes are a tough sell, but she also knows that every single word and number in the presentation is there to help the company survive. And they all want the company to do well, right? If not that, then why does she work so hard? If it's not for the sake of more sales then what is it for?

However, despite the highly polished speech and visuals, the clarity of the numbers and the conviction of her voice and gestures, the presentation failed to impress. Sure enough everyone got the point and they agreed on its usefulness. But something significant was missing. Main comments afterwards, mostly during the informal dinner, were that the strategy is spotless but that necessary changes do not feel engaging. Even the CEO, talking to her briefly at the dinner, remarked that the room was rather cold during and after the presentation. Next morning, she did what she always has done in such situations: rolled up her sleeves and started working even harder to make the plan come true. Implementation of the first phase shows regional directors performing well but not as well as she would like. Some are doing better than others, but overall they are not as supercharged in changing their ways as she would expect them to be in such challenging times. In all honesty, even she is not as enthusiastic as she imagined she would be at the beginning of the assignment. Something is indeed missing for her as well. Learning and change are slow and painful for everyone. After a few months she steps down. She no longer has the inner push to get her going and to over-perform as in the past. What happened? The task was clear, the numbers were strong and everything was in order for success. Everything apart from the brain.

Let's be honest: great leadership goes hand in hand with great performance. Our ability to perform with excellence and influence others to do the same does not depend only on our traditional view of performance, namely on a mix of skills, competencies, time, resources and guidance. It also depends on the brain's ability to recognize the significance of the task in hand and to produce the commitment, endurance and changes needed for success, regardless of barriers, failures and shortcomings. This ability comes from a remarkable brain characteristic which surprises everyone coming across it for the first time. Apparently, our brain can change a lot during our lives and this change can determine the outcome of our efforts in a critical and decisive way.

The ever-changing brain

The brain never stays the same. It constantly changes. The brain is a dynamic system of neurons that are constantly moving. This remarkable fact has taken neuroscience by storm and cancelled long-held beliefs that all brain changes happen only when we are very young. As reported by psychologist and neuroscientist Elaine Fox (2013) in her book titled *Rainy Brain, Sunny Brain: How to retrain your brain to overcome pessimism and achieve a more positive outlook*, it used to be the mainstream notion up to the 1980s that the brain is a biologically static organ, meaning that its structure and nature do not change inherently during our adult lives. So, more or less, we got stuck with our brain for our whole life. Our duties were first, to make sure we did not damage it in any way and second, that we used it as best we could. Although it could always get worse, through accident or illness, it could never get better. The upper limits of our brain were set in stone since we could not drastically improve its learning abilities and performance. This is why IQ tests were so important. We had a stable brain capacity that we could measure and figure out its potential. This cannot be further from the truth.

For more than two decades now we have known that the brain is everything but stable. Our brain is plastic, meaning that it has the ability to rewire itself, creating new synapses between its neurons based on what is happening inside and outside of us. Our understanding of neuroplasticity, as this ability is called, has altered the way in which we view the brain and our capacities in life. The way we use our brain can actually change it; killing the previous view that the matter in our brains is the rigid 'hardware' and our thoughts are simply the 'software'. Neuroplasticity shows that our 'software' can change the 'hardware', which is difficult to comprehend since in order to improve our smart devices we need to purchase hardware upgrades and technical plug-ins. Instead, concerning our brain, the 'hardware' upgrade can happen by upgrading the 'software' or by changing our behaviour. As described nicely by Chopra and Tanzi (2013) in their book *Super Brain*:

> Neuroplasticity is better than mind over matter. It's mind turning into matter as your thoughts create new neural growth... Our brains are incredibly resilient; the marvelous process of neuroplasticity gives you the capability, in your thoughts, feelings, and actions, to develop in any direction you choose.

The remarkable ability of our brain to rewire connections between its neurons has led to the now famous phrase 'cells that fire together, wire together', which actually suggests that the more we use particular neural pathways the more these pathways will be strengthened. Although the advantages of such neural activity are numerous, it also has its downsides. First, connections between neurons can become too strong, leaving little room for flexibility and change and second, areas that are not used can grow very weak. 'Don't use it and you'll lose it' is another famous phrase on neuroplasticity. It practically means that if you do not adapt and evolve, you do not just remain the same as before, but you actually deteriorate your neurons and their connections, becoming less able to perform some tasks than before. Eventually, you become cognitively weaker and professionally more vulnerable.

In an unpredictable and continuously changing environment where organizations need to keep up with changes coming from everywhere, leaders' brains need to constantly adapt and evolve. Neuroplasticity can be our best friend or worst enemy depending on how we use our brain. Neuroplasticity works in a leader's advantage through the process of constant learning, since learning is the main fuel for neurons to create new connections and paths. Bruce Hood (2014), a renowned British psychologist, clearly stated in his book *The Domesticated Brain* that there is positive plasticity in our brains, even in our adulthood, as far as we constantly learn throughout our lives. Learning is the key to plasticity as it is the key to performance for modern leaders. As Indra Nooyi, Chairman and CEO of *Pepsico*, advised professionals in *Fortune* (2014) magazine:

> Never stop learning. Whether you're an entry-level employee or a CEO, you don't know it all. Admitting this is not a sign of weakness. The strongest leaders are those who are lifelong students.

A plastic brain that never stops learning is a brain with improved cognitive skills such as better memory and enhanced attention. So it plays an instrumental role in developing our neocortex capacities and this is the main reason it appears within the first pillar of our model. But how do we keep ourselves motivated to constantly learn, adapt and evolve? Where do we find the inspiration and the drive to go first where no one has been before and to take others with us? Most importantly, how do we keep our brain alive, firing and growing through challenges and even through disappointments? First of all, a leader with an active plastic brain is a leader with a strong purpose.

Purpose above all

The main problem with the new global sales director in the opening to this chapter was that she did not have the right attitude in her new role. Although she managed the tangible elements of her job, such as numbers, strategies and presentations, well her performance was not driven by a higher purpose. And the higher you climb the organizational ladder, the more you need a great purpose to drive you, guide you and make those around you follow you. This is something that no modern leader can ignore anymore. Instead of focusing on sales as the ultimate goal and main motivator both for herself and the others, she should have targeted different purposes activating other brain regions.

In one of the most-watched TED speeches online, Simon Sinek explained the simple biology of purpose. Starting from inside the brain and using the famous cases of Apple, the Wright brothers and Martin Luther King, he famously argued that the most powerful motivation to do something comes from deep inside the brain. These older brain structures are moved by passion deriving from answering convincingly 'why' questions. In contrast, most companies and people try their best to answer to 'what' and 'how' questions, forgetting that the true power lies not in the complicated explanations of products, process and procedures but in big ideals. A great, and preferably long-term, purpose can help people around you to see the direction and understand better the decisions made and the decisions that will be made. Sinek (2010), in his book *Start with Why*, described his golden circle that holds in its centre the mighty why question. If you are driven by a powerful 'why', which is related to a higher purpose, then you will put in the hours, commitment and even the influence needed to achieve your goals. If you are driven by 'what' and 'how' questions you will achieve very little in difficult times and bring nobody with you. 'Why' brings passion and connections from deep inside the brain, the areas that really drive behaviour, while 'what' brings rationalization, analysis and very little movement. This claim is remarkably similar to Daniel Pink's assertion, in his book *Drive* (2009), and in his own famous TED speech on the puzzle of motivation, that contemporary problems require creative problem solving that can only come from intrinsic motivation. If you are after money and sales for the sake of it you will not be able to reach the more elusive but more effective brain regions that bring solutions to unexpected problems. Money and sales are part of what have been called extrinsic rewards (Buchanan and Hunczynsky, 2010), which are measured outcomes set and controlled by others. This extrinsic motivation can only work with straightforward, routine-based problems.

Intrinsic motivation frees your brain from the pressure of immediate and short-sighted rewards and allows you to perform better than before. Therefore, this motivation works better for difficult, challenging and complex targets that require our creativity and complex thinking as well. This seems to be the case in an experiment that we have conducted with our students. We separated the students into two groups. Each group had the same target. They needed to enter a room one by one and complete a puzzle, which was the same for both groups. In the first group we asked them to complete the puzzle without promising any reward, saying that we will just count the time that each needs in order to fulfil the target. We let them know that we are grateful to them and will acknowledge their participation in this experiment, which will give us extremely valuable knowledge. In the second group, we asked them to do the puzzle, ensuring a bit of competition among them by promising a specific monetary reward to the three fastest students. What do you think happened? According to traditional (rational) management jargon, the pay-for-performance situation (second group) works better than the non-extrinsic reward situation (first group). The results show the complete opposite. The first group of students was 3.3 minutes faster, on average, than the second group. We have repeated this experiment many times and, in the majority of cases, we observed that the promise of a specific extrinsic reward worked as an obstacle in motivating their brains to think in a creative manner. In contrast, when we facilitated the emergence of intrinsic rewards students performed better with challenging targets. We have also observed that intrinsic rewards are enhanced with a specific overall purpose. Thus, intrinsic motivation needs a purpose, and purpose apparently starts with 'why', deep inside the brain.

Purpose, as a key leadership characteristic, was highlighted much earlier though by UK-based consultant Nikos Mourkogiannis in his 2006 book *Purpose: The starting point of great companies*. Mourkogiannis claimed that in order to become a great leader one cannot afford to be out of sync with those around them with the purpose that motivates them. So, aspiring leaders need to develop a clear sense of their own moral compass to help them guide themselves and others in future career-defining and life-defining decisions. According to him there are four main sources of higher purpose in organizations:

- *Discovery*. For people driven by discovery, life is an adventure. The continuous search for new discoveries pushes people to make strong promises and to do their best to deliver on them. Tradition, familiarity and current constraints are the worst enemies of discovery. People and companies guided by this purpose will always go the extra mile to pursue their aims and especially to be the first to serve a new, untapped market need.

- *Excellence*. For those driven by excellence, life is an art form. Their actions do not derive so much from customer demand but from an obsession over the inherent qualities of the product or service. Their output has to be the best possible, the best available; in one word, perfect. Like artists, their high standards are their guiding light and are non-negotiable. Lowering standards in processes and outcomes is considered as compromise and even defeat.

- *Altruism.* For those employing altruism as their purpose, life is simply about supporting others. It refers to the personal or institutionalized instinct for providing help to our communities. This caring attitude can be towards customers, employees, suppliers, towns and even the environment as a whole. It is actually about having the well-being of those around you as a top priority and your own selfish needs as a second or third.
- *Heroism.* Those believing in heroism believe that single individuals, or single companies, can and should change the course of history and influence for the better the lives of everybody else. They feel that they possess special skills and they should use them fully to achieve the highest possible results, defying the status quo at any cost. Heroism is a highly competitive attitude, but also one of sacrificing one's personal life for the ultimate reason of being on this planet: to develop that product, to build this company, to close that deal, to solve this problem.

In our experience, purpose is the starting point of a great leader, a great colleague, a great friend. Again and again we meet managers and directors that are strong in skills and knowledge, they have clear career aspirations, they even have a good track record of results, but lack the internal fire to bring themselves and those around them to a new level. We usually advise those individuals that whatever brought them there will not necessarily make them successful at a higher level in the organization. A move from the tangible to the intangible, from just targeting more sales to enhancing their creativity and complex thinking, is a prerequisite for them to endure the challenges of steering companies and top-level teams. It is needed to keep them going, keep them learning, keep them connecting and keep their brain improving. A higher purpose is not a leadership luxury but a leadership prerequisite.

How to spot the wrong purpose

We always start our coaching sessions by enquiring about the person's purpose. Through a dialectic process of exploring motives and aspirations we guide our clients to perform a simple diagnostic of their own purpose and of those of people around them. This diagnostic consists of the following four questions in that specific order:

1 *Extrinsic or intrinsic?* If the individual is driven by short-term, overly focused, egoistic and materialistic motives then they are not on the

right path. Extrinsic rewards that follow a narrow-minded 'if-then' logic ('if I do this then I get this'), do not allow for the full potential of a true purpose to be realized. A true purpose needs to derive from higher ideals such as positive contribution to the lives of the people around us, and from deeper emotions such as love and caring. We help them discover by themselves the potential 'why' of their actions that often is hidden deep inside them.

2 *Complicated or straightforward?* If the individual uses too many words to describe his or her purpose then something is wrong. Words do not work in favour here simply because the deeper brain structures that are responsible for commitment, passion and behaviour do not have language capacity. The more words used, the stronger the indication that the executive part of the brain in the neocortex has taken control and thus not much motivation is expected. Great purposes are simple to explain and powerful in moving people. Of course, if the individual replies in a fast and straightforward way but mentions extrinsic drivers then we start again. We help them to express their purpose in a way that everybody can easily understand it.

3 *Forced or natural?* When an individual uses corporate language and/or just repeats the mission, vision and values of the company then we have a good reason to be concerned. A higher purpose is genuine when it comes naturally to the individual and it is not imposed in any way. Even when someone is aligned with most of the organization values this does not mean that a carbon copy of corporate materials will be produced during our enquiry. When this happens we suspect that our client is just hiding the real purpose behind the official one. After further discussion and some homework people usually expose their drivers and then the real work starts.

4 *Low or high share potential?* Here we explore whether the higher purpose is easily transmittable to other people within and outside the organization or not. If yes, this means that this purpose will not only serve as the leader's guiding light but it will engulf internal teams and external customers, suppliers and partners as well. Higher purposes that go viral through a company's people, products and communications have changed not just individuals, but whole industries and societies too. If the purpose is not easily transmittable to others and cannot inspire them then we need to review either the purpose itself or the environment, or both.

> **Action box**
>
> - *Individually.* Go through the four questions in the 'How to spot the wrong purpose' box. Make notes. After you answer each question consider your notes based on the discussion in this chapter. Is your purpose in accordance with best practice? If not, what can you do to improve it? Rewrite your answers until you reach a satisfactory result. You can use the Mourkogiannis model to help you define or redefine your purpose.
>
> - *As a team.* You can do this exercise with your team as well, but only if your search for the right purpose is genuine. This will lead to other people being open and sharing their inner thoughts and feelings. Identifying different higher purposes within the team is not a problem. In such cases, discuss with your team how your different purposes can help you to support each other in achieving your organizational objectives. Shared inspiration is what we aim for. We suggest proceeding with the team task after you have spotted and defined your purpose on an individual basis.
>
> Purpose motivates you from the inside out to always move forward and upwards, driven by an unlimited passion and a higher principle. But purpose will never help you do anything extraordinary if it is not accompanied by the brain state of top performance. This state is called the *flow* and on your way to great leadership you need to embrace it fully and enthusiastically.

Flow to greatness

The brain does not perform optimally by default. It can do so under specific circumstances that can be observed and analysed. We can then induce this optimized brain state and achieve what we're aiming for, accelerate our learning and develop expertise much faster than expected. This realization has revolutionized the way we see performance. It should also make neuroscience the closest 'friend' of all professionals around the globe.

When was the last time you felt and performed at your best? How often and how easily do you find yourself in perfect concentration and in absolute control of the moment? When were you able to absorb and comprehend information fast and deliver effective solutions that made you and your team proud? In short, when and how are you getting in the flow? What most people call 'being in the zone', in neuroscience is called the flow. It describes the moments of total absorption that lead to an optimal state of consciousness. Optimal state of consciousness means best possible performance, output and well-being. Mihaly Csikszentmihalyi (2008), the famous Hungarian-American psychologist who first methodically studied this state, devoting his life to it, fittingly stated that:

> [T]he best moments in our lives are not the passive, receptive, relaxing times... The best moments usually occur when a person's body or mind is stretched to the limit in a voluntary effort to accomplish something difficult and worthwhile.

Almost everyone can recall a moment where, in high adversity, they found the inner strength to do their best, feeling fully empowered, focused and content. In such situations we experience a kind of alternative reality where time is altered, going too fast or too slow, and the sense of ourselves merges with its surroundings. In short, we are unstoppable! Such a state is critical for leaders to perform amazingly well when needed the most and create followers who will believe in them whatever the situation.

The future belongs to the flow and many remarkable projects are taking the concept further. The neuroscientist and entrepreneur Chris Berka and her team have created a simple-to-use device that you can wear on your head to measure the two key brainwaves involved in the flow, alpha and theta (Advanced Brain Monitoring, 2015). According to the *American Heritage Dictionary of the English Language* (2011), alpha waves can be consider as a pattern of smooth, regular electrical oscillations in the human brain that occur when a person is awake and relaxed and they have a frequency of 8 to 13 hertz. According to the same source, theta waves have a frequency of 4 to 8 hertz, and they are recorded chiefly in the hippocampus of carnivorous mammals when they are alert or aroused. Alpha waves produce a state of meditative concentration and theta waves extreme relaxation. Imagine measuring your brain waves while you consciously try to tune them for flow. As soon as you achieve the right measurement in alpha and theta brain waves through relaxation and concentration, you start dealing with a difficult and crucial situation. You are now in optimized performance. According to Berka, such technologies will allow you to learn, be more creative and master new skills faster than ever. Furthermore, Steven Kotler (2014) and his Flow Genome Project are committed to uncovering the neurobiological aspects of the flow. By looking at the chemical, electrical and structural changes in our brains before, during and after the flow he aims at providing solutions for people to achieve it more often.

In our experience the key internal and external conditions for achieving the flow are the following:

Internal

- *Higher purpose*. There is no flow without feeling that what you do serves your personal higher purpose. The more you are genuinely, actively and passionately motivated by your purpose the easier it is to get into the flow.
- *Advanced expertise*. Although flow accelerates learning and helps you develop personally even further, it cannot be achieved if you lack the basic skills of your trade. You need to be good to become excellent and perform with excellence. And excellence can be achieved after practice. In their study Howe *et al* (1998) have argued that a talent (like leadership) can become a critical skill mainly because of long and hard work.
- *Low anxiety*. The higher the level of anxiety, the lower the possibility of getting into the flow. Bad stress and too much pressure are the worst enemies of getting into the optimal state of flow because they do not allow the winning combination of meditative concentration, strong emotional rewards and intrinsic motivation to manifest simultaneously.

External

- *New challenge*. You will probably not go into the flow if you are facing a routine problem. Routine problems put the brain into autopilot and habits are called to finish the job. No need for wasting additional brain energy if the challenge does not include novel elements that make it exciting and engaging.
- *Some control*. It will also be very difficult to get into the flow if you have absolutely no control over the situation at hand. Without empowerment and the ability to take decisions freely to solve the problem there is little space for finding the inner strength to become the best version of yourself. A sense of autonomy and empowered contribution go hand in hand with the ability to achieve flow.
- *Increased significance*. The more important the situation the easier it is to get into the flow. Problems with limited implications, regardless of how surprising and novel they are, cannot create the sense of urgency and the feeling of increased impact that can bring about flow naturally and convincingly. We get into the flow when it is most needed.

Whatever the next step is in understanding flow better, one thing is certain: our schools, universities, institutions and organizations should move towards becoming flow-friendly organizations where people get into the flow individually and collectively to achieve, create and enjoy more. The real question here is: are you becoming a flow leader? That is, are you working on getting yourself into the flow and also enabling others to do the same? This is the fastest and surest way to brain-based leadership.

Creativity killed the competition

We have already discussed in the previous chapter the importance of asking questions for keeping a clear mind. However, being curious is not just about maintaining a good view of the world around us. It is also crucial for being creative and bringing new solutions to new problems. In dynamic and unpredictable environments where rapid changes are the norm, creativity emerges as the number one skill for executives and leaders. It also seems that creativity is the main (if not the only) way to respond to the increased complexity that surrounds us. So, we need to move fast, away from old, obsolete and risky sayings such as 'curiosity killed the cat'. The only thing that curiosity, and creativity, can really kill is your competition!

An IBM survey of more than 1,500 CEOs from 60 countries and 33 industries around the globe in 2010 found that creativity is considered *the* essential skill for effectively navigating the increasingly complex business environment. CEOs in this study placed creativity as the top required skill, above more traditional business skills such as rigor, management discipline, integrity and vision. This is a remarkable result since creativity is traditionally considered as a skill reserved for specific professions such as artists, advertising gurus and inventors. Indeed, creativity was rarely a core subject in business studies, seminars and trainings outside marketing communications subjects and innovation courses. A study conducted by Ros Taylor (2013) in the United Kingdom for her book *Creativity at Work: Supercharge your brain and make your ideas stick* across various industries with 100 executive interviews revealed that 70 per cent believed that creativity was less compatible with the workplace and more compatible with arts. Her subsequent survey of 1,000 professionals across the country showed the lowest scores in the questions 'creativity is at the top of my organization's agenda' and 'everyone [in my organization] systematically uses a creative tool or technique'. Regardless of how badly CEOs want to have it in their companies, creativity is not part of the DNA of traditional business. The worldwide obsession though with continuous innovation at all levels has brought about creative thinking, as we have argued above, as the main weapon for leaders to deal successfully with complexity. But where do creativity and creative problem solving come from?

The part of the brain that brings the most insightful new ideas is not the part of the brain that does most of the analysis and rational thinking. Evangelia Chrysikou (2014), an Assistant Professor of Psychology at the University of Kansas, summarizing the research on the neuroscience of idea generation, has argued that lower activity in the executive part of our brain is crucial for creativity. Hypofrontality, which occurs when there is diminishing activity in the prefrontal cortex, leads to fewer constraints, lower attentional focus and the abandonment of rules. The mind becomes more open and new ideas emerge more freely. The quieting down of the executive brain is positively associated with a higher creative output. The less we think about a problem, the more creatively we can solve it. Evidence also comes directly from the industry. Mike Byrne, founding partner and global chief creative officer of Anomaly, an unconventional marketing agency with offices in three continents, revealed in *Entrepreneur* (2015) magazine:

> I give every idea time to 'bake'.... The best baking always comes when I am doing something else. For some reason it opens up my thinking. I get clarity. I can solve a problem quicker because the baking helps me get to the answer quicker.

It is not uncommon for executives and other professionals who rely on creativity to refer to running, showering, taking a walk and even 'sleeping on it' as great strategies in reaching creative solutions to new challenges. Executive MBA students and a few corporate clients of ours have expressed very similar opinions. This is not to say that not thinking actively about a problem is ignoring it. 'Baking' an idea, as Byrne put it, is not running away from it. On the contrary, quieting our executive brain *after* learning as much as we can about a problem, not before, is crucial for the rest of the brain to offer us the proper relevant solution. Nevertheless, it seems that the more we squeeze our brains to produce ideas by looking again and again over data, graphs and spreadsheets, the fewer creative solutions we will be able to find. This counterintuitive proposal goes against the well-established practice of many companies and managers where more power is allocated to the executive part of the brain whenever a difficult and complex problem arises. Paradoxically enough, it is when creativity is needed the most that we minimize our brain capacity from producing it.

The pioneering work of Jennifer Mueller, from the University of San Diego, on creativity in organizations supports the above. Her work suggests that creativity is inhibited greatly when executives think in usual business terms. In her popular and highly cited 2012 paper 'The bias against creativity: Why people desire yet reject creative ideas', she and her associates concluded that even if organizations claim to promote creativity and innovation they actually stop it when they require their staff to come up with the best, most appropriate and readily feasible solution. Organizations' inherent aversion to uncertainty and its avoidance will quickly kill any really creative and novel idea simply because managers will be in favour of safer options on the brainstorming board. Thus, there is a bias against creativity: although we declare we want it, we do not really have the mindset and processes in place to accept it. We want it but we actually try to avoid it.

Implications of the above study are immense since there is an urgent need to change the way we perform idea generation workshops and our filtering process, because currently it filters out most of the actual creative ideas. Too much thinking will inevitably bring into the picture the executive brain and its eagerness to over-analyse, evaluate risks, calculate consequences and predict future outcomes using what is already known to it. It will eventually regress to the familiar, to the risk-free, to the expected. The executive brain is not the right brain to bring and even judge new ideas in their infancy stage. As a result, we need to quieten it down in order to allow all-powerful insights to surface.

In this respect, we suggest the following six-step approach, based on available science and our experience, to foster innovation and allow creativity to flourish in your organization:

Step 1. Understanding. Here you need to become as familiar with the problem as possible. You need to do the research, read the key data, look at the main graphs, discuss with all the key people and consider all possible angles. In this step you use your analytical brain to get important information which, together with your personal experience, will provide the raw material for your insightful brain to do its job. The team should do the same.

Step 2. Exploration. When you are trying to come up with ideas and discuss them with your team keep your mind as open as possible. Too much analysis and criticism at this stage does more harm than good.
2a. Individual exploration. Thinking without thinking is important: keep your brain not on the problem but on the background. Shift it into the foreground when needed to discuss it with the team and/or to consider new evidence. 'Baking' the problem, 'sleeping on it', running or exercising to free up valuable brain space, listening to music, and shifting your brain to higher thinking can help immensely.

2b. Group exploration. Discuss in a playful and enjoyable way even if the problem is very serious. Try to keep yourself and the team away from the stress of overplaying responsibility, failure and disaster. Take breaks often, divert team attention when the discussion is over-heated, take a walk together. A smiling, engaging and purposeful team is a creative team almost every time.

Step 3. Decision making. When the preferred idea (or ideas) surfaces, instead of rushing off to implement it, consider it carefully. Does it really address the problem? Have we considered all angles? Did we make sure that exhaustion, fear, feasibility, easiness and risk-avoidance did not make us reject the real creative options? Ask for an opinion from trusted sources outside the company because they usually see problems and solutions without the team's internal biases.

Step 4. Implementation. Putting the idea to work nowadays requires as much creativity as its conception. Planning, allocation of resources and coordination of activities are no longer such straightforward tasks as they used to be. Constantly reviewing the process by getting personally involved, always asking the right questions and passionately keeping the purpose alive for all can do miracles for bringing innovative solutions to life.

Step 5. Results. Success or failure is based on the final outcome. Measuring the obvious requires less creativity than measuring what matters. Examine all possible ways to measure the impact of the implemented solutions. Include as many metrics as possible – both quantitative *and* qualitative. Celebrate successes in order to cement the creative approach to business problem solving. Deal with failures creatively since there are always important lessons to be learned.

Step 6. Configuration. Embracing creative thinking opportunistically when trying to resolve a specific problem is fine. But the real long-term solution is to develop a culture and structure that constantly promotes creativity. Such an environment is characterized by playfulness, empowerment, constant questioning, appropriate tools and a feeling of increased purpose in everything we do. As the famous experiment by Kempermann *et al* (1997) on mice showed, when a group spends time in an engaging, lively and fun environment it creates three times more neurons in specific brain areas than the group that spends the same time in comfortable but empty settings. Transferred to people, such drastic neurogenesis can form the basis for great achievements within companies. Modern innovation and creativity experts, such as Stanford's Tina Seelig (2012), never fail to emphasize the importance of the right environment and culture for creativity, with others, such as Ros Taylor of *Creativity at Work* (2013) highlighting the need for specific techniques, tools and procedures for ensuring that a creative process will always be applied.

> **Boost your brain**
>
> Observe closely the process of idea generation and decision making next time your team holds a brainstorming session. Did the team use their brains appropriately to promote creativity or did they allow their prefrontal cortexes to lower risk and choose familiarity? If this is the case, then come up with (creative) ideas of how to approach the whole process anew to increase the creative output and promote those ideas within the team.
>
> Option 1. In the next brainstorming session choose a provocative idea from the board that works well for solving the problem but that is not popular with the team because of feasibility issues. Ask everyone to pretend for a moment that this is the chosen solution. Observe how the team behaves now and facilitate the discussion accordingly. This will shift their attention from dealing with uncertainty to dealing with an exciting new project without feeling the weight of responsibility. Creativity will be released!
>
> Option 2. In the next brainstorming session ask participants to come up with ideas quickly, trying not to evaluate them at all in this phase. Write all these ideas down and then ask everyone to mark each idea from 1 to 5 based on two variables: actual problem solving and feasibility/comfort. Discuss any discrepancies and reinforce the need for a solutions-oriented view.
>
> Option 3. Ask team members to discuss ideas by advocating and supporting *only* their ideas, and avoiding criticizing the ideas of others. Help them to release creativity by focusing on the positive rather than the negative aspects of decisions.

Creativity is neuroplasticity at its best. Identifying new solutions to new (or old) problems, learning new ways, exploring unexpected insights and challenging the status quo are enabling new neural pathways to regularly be built inside our brains and keep us effective and competitive for a long time to come.

If memory serves me right

We have observed, and we are sure you have too, that in organizations people with great memory are very much admired. Actually they are admired everywhere. When someone's memory is readily available to provide a fact, an event, a name or a date from the past in order to help a discussion progress, this impresses those around them. Such experiences of ours, when working with organizations around the world, lead us to believe that improving your memory is a key part in calibrating your brain for better leadership. And we are not the only ones who believe this. A poll by *Scientific American Mind* in 2014 revealed that the majority of readers participating in the survey prioritized cracking cognition on their wish list of an enhanced brain over building character and even curing disease. Within this category the most popular ability they wanted to master was memory, topping the whole survey with 40 per cent of the total votes. Joshua Foer (2012), author of the book *Moonwalking with Einstein: The art and science of remembering everything*, describes it like this:

> The people whose intellects I most admire always seem to have a fitting anecdote or germane fact at the ready. They are able to reach out across the breadth of their learning and pluck from distant patches... memory and intelligence do seem to go hand in hand, like a muscular frame and an athletic disposition.

And it is not just a powerful impression that good memory helps leaders achieve. It also improves decision making and thus 'thinking' in our brain adaptive leadership model. The complex neural process of taking a decision depends partially but crucially on memory. Memory is responsible for bringing back knowledge and past information in order to deliberate effectively on possible courses of action during the decision-making process, according to Antoine Bechara (2011), Professor of Psychology at the University of Southern California. In essence, a missing memory can lead to repeating mistakes and taking wrong decisions that could have been avoided. How many times have you participated in a meeting where the decision was swayed by an important fact or past experience retrieved from memory by a team member? Everyone's appreciation for this team member must have at least doubled in that meeting!

Our memory is largely categorized in the short term and long term. Hippocampus, the seahorse-like part of our brains located right under the neocortex and in the limbic system, plays the key role in short-term memory, consolidating memories from the short- to the long-term one and in spatial navigation. People with a damaged hippocampus have both memory and orientation problems. Memories though are stored in the synaptic connections of neurons in our neocortex. In order to be able to remember something in the long term we need to fight the natural process of memory degeneration by creating a dynamic and repetitive process between our hippocampus and our neocortex. Information that is not revisited tends to get forgotten since the synapses that hold this memory weaken over time. Donald G Mackay (2014), a Professor of Psychology at UCLA, argues that:

> Just as a builder can make a new structure or repair a damaged one, so could the hippocampus craft new memories to replace those that have been degraded... Such rebuilding might occur whenever someone reencounters a forgotten word or a personal anecdote from the past.

This is the way by which more recent and frequent exposure to a piece of information can restore a fragmented memory and minimize memory loss rate. The process for this being done as suggested by Mackay is the following:

- first, information is captured by our senses;
- second, it is transmitted to the relevant cerebral part for processing (let's say the visual cortex if it is a visual stimulus);
- third, this information travels to Borca's area in the neocortex that stores memories of words;
- fourth, the person cannot retrieve this information due to time degeneration;
- fifth, the person is exposed again to the same information;
- sixth, the new stimulus is transferred to the hippocampus that interacts with the neocortex to recreate the forgotten memory; and
- seventh, the reconstituted memory is stored again but with new strength in the neocortex.

Practically, we need to remind ourselves of important information by revisiting original or indirect sources and thus making sure we rebuild and re-strengthen faded memories. Do not leave this to luck and nature!

Another way of powering up our capacity for stronger long-term memory is emotions. By simply caring personally and engaging more in the activities you are involved in, increased emotions will help you retain more information. Feeling detached and bored will have a negative effect on what you will remember from a conversation, meeting or activity. This is because by engaging ourselves actively in a situation we care a lot about, we develop autobiographical memories. This means memories that are close to our core self. Emotions in such a case act as a glue to make sure we remember important information, both good and bad. As psychologist Stephan Hamann (2001) suggested, emotional arousal influences memory both during memory encoding, which means attention and elaboration of the information, and also in memory consolidation. In essence, memories supercharged by strong emotions are reactivated more, remembered better and attract more attention to them. It is imperative then to double-check your motivation, involvement and emotional attachment to your work and to working situations. The more personally you take your job, the more memorable your working moments will become. In a similar manner, the more personally you take your role as leader the more memorable your leadership attitudes will become, helping you to repeat them or change them. A strong sense of purpose comes in handy here as well.

The third way for improving your memory is by utilizing associations. This technique is used by those competing in memory championships around the world and it is based on the understanding that the strongest and most varied associations of a piece of information in our neocortex, the easier it will be for us to store it and retrieve it. Memory is a spatial-related cognitive activity. It is crucial for each new piece of information that needs to be stored effectively to be involved in an action that will make it distinctive and irresistible to memory. As eloquently described in Foer's (2012) book about the art of memory, this is because information needs to take as much space as possible within our cortex by creating as many and as strong synapses as possible. A mundane event can become extremely entertaining if we associate it with fun elements that we completely make up and attach to it on the spot. For example, if you struggle to remember the job title of a new acquaintance think of a friend of yours that has the same job title and imagine him riding uphill on a lama or a donkey with the new acquaintance. This unusual and amusing mental image will increase associations in your brain and will help you remember it more easily.

Action box

Based on the ancient method of the poet Simonides of Ceos called the 'Memory Palace', memories need to be stored in a carefully constructed mental image of a familiar structure (house, road, neighbourhood etc) by attaching to each piece of information a specific location within this structure.

This exercise needs to take place in a silent location. First, choose five new pieces of information you want to memorize. Then close your eyes and mentally walk yourself through your office (or home). See every corridor, corner, desk and meeting room in your floor in all possible detail but empty of people and noise. After you finish and you are satisfied with the clarity of your memory palace, start mentally walking around it again, placing the five items (images, names, notions, words, whatever they are) in a specific order in different locations within the office. You have to vividly put them down in each distinctive place. To increase memorability you need to also create a unique activity for each item in each location that involves a verb and a person or an item actually doing something. Remember, the more unorthodox, entertaining and edgy the act, the more successful you will be in storing the memory.

After you finish, take a small break and then relax, close your eyes and try to revisit your memory palace, mentally walking around it and finding the items you wanted to remember stored in the places you left them. Practice makes perfect, so the more you play this game the better you'll become. Just make sure you mentally clean your palace before you start storing new items.

Memory has to serve a leader well. Improved memory will lead to better decisions, more meaningful conversations and overall increased performance. Most importantly, it will dazzle the people around you and will help you enhance your charisma and influence effect. Furthermore, memory can increase your brain's ability to adapt, which is equally critical for leaders nowadays.

Adapt, bet and grow

The *Economist* publishes annually its most important predictions for the year to come. The esteemed 'The World in [Year]' publication includes a number of predictions in economics, business, politics and technology for the next year. In its 25th year special edition 'The World in 2011', the prediction for the country Libya in the Middle East and Africa section stated:

> Muammar Qaddafi has held power for 40 years and will certainly complete 41. By suppressing opponents and undermining rivals, he has removed all significant threats to his rule; the only credible successor is his own son, Saif al-Islam.

Knowing that 2011 was the year that General Qaddafi and his family were ousted from power and Qaddafi himself was killed by rebels, one can only be startled by the confidence in that particular prediction. We've seen in the previous chapter that such an attitude fits better the profile of a hedgehog than of a fox. It also portrays a simple but powerful assertion for the world we live in and the corresponding mindset leaders need to adopt to survive. The level of complexity we experience today, based on the high number of interrelated forces simultaneously shaping all aspects of our lives, supports a very different approach in decision making for leaders of all types. From the out-dated approach of choosing one single and rigid possible future and behaving accordingly we need to move to a more flexible, open and careful approach of choosing multiple ways forward and pursuing them with not necessarily equal, but definitely critical attention. This new approach is consistent with the popular writings of modern philosopher Nassim Nicholas Taleb (2007, 2012) on randomness, black swans and anti-fragility. In particular, Taleb claims that randomness and unpredicted events are far from being under our control and ability to foresee. Stability, routine, incremental and planned improvement is something that humans prefer and target rather than radical and sudden change. However, this situation, given that unpredicted events (black swans) will come sooner or later, make us more fragile and sensitive to their impact. In contrast, our constant exposure to small variations (avoiding from time to time routinized processes of actions) can make us potentially more resilient, or, as Taleb argues, *antifragile*. This approach calls for what has come to be known as small or purposeful betting. This is the best cognitive strategy we know of and apply for steering people and organizations through the dangerous waters of unpredictability.

With our digital selves, mobile devices and social media dominating the collective global conversation, with political uncertainties and conflicts constantly bursting around the world and with a start-up culture challenging and dethroning industrial behemoths on a daily basis, leaders have little room to be certain about the future. This creates a serious issue with strategy, personal and organizational, since strategy, a critical aspect of leaders' thinking, traditionally relied on predictions of, and decisions about, the future. If the future is impossible to foresee how can we make any solid decision on investments, resource allocation, preparation and activity planning? By employing the mindset of the purposive bets we can deal with our chaotic reality more effectively and at the same time lower the overall risk of failure.

Purposeful bets are defined by author and consultant Frans Johansson (2012) as the actions we take regardless of our ignorance of their effectiveness, and by entrepreneur and author Peter Sims (2011) as actions of low risk intended to unearth and try ideas out. Since high uncertainty defines all areas of human endeavour and we cannot stay idle, it is advisable to take as many diverse actions as possible at the same time, investing in – or betting on – their success. This will eliminate the possibility of a fatal blow if one or some of them do not work out. The leadership mindset deriving from this approach is one of braveness, hyperactivity, connectedness and optimism. The executive part of our brain does its usual analysis, forecasting and decision making by quietening the amygdala's fear of not surviving. If many small bets fail all is not lost. However, if just one succeeds it could make a huge difference in the overall outcome.

The five key steps in placing purposive bets, which we also have been actively applying, according to Johansson (2012) are:

1 *Place many bets*. Playing in only one arena and envisioning just a single possible future can be catastrophic. Engaging in multiple activities simultaneously and always discovering and following new ways to pursue goals are necessary for success. For example, when faced with a specific challenge, do not search for the one and only 'golden solution' but engage in exploring various alternatives. We recommend choosing a number of actions that can co-exist in implementation rather than being mutually exclusive to each other.

2 *Minimize the size of the bets*. The main characteristic of bets is that they are low risk. This means that although we do commit resources and time to implement them we do not base the financial and strategic survival of our organization on them individually but collectively. We have observed that committing all resources in one solution can be fatal both for a company and for managers' careers. We need to move away from the all-powerful magic idea that can solve a problem instantly (but expensively too) to a manageable portfolio of more cost-effective solutions.

3 *Take the smallest executable step*. Since bets are used to examine ideas in practice and many of them simultaneously, they should be approached with caution. A methodical approach should be adopted where every task leads to the next only when the validity of the previous step is established. Allocating different steps to different people, using external resources (researchers, consultants) to help verify the most important of them, and using your closest team to evaluate progress and approve next steps, are key actions that we suggest to our clients.

4 *Calculate affordable loss*. Instead of analysing the possible return on investment (ROI) of every bet in order to decide whether to proceed

with it or not, give the go ahead based on how much it is acceptable to lose in each one. ROI can stop you pursuing activities even with very low investment requirements without really ensuring that the ones chosen will actually deliver as expected. So, it is better to stop having expectations that are anyhow highly uncertain and focus on getting bets off the ground by investing as little as possible in each of them.

5 *Use passion as fuel.* Constantly placing numerous bets and following them closely, even witnessing many of them dying out, requires a very strong commitment. As already discussed in this chapter a higher purpose, passion and intrinsic motivation can keep you going, learning from the failed bets fast and using every possible success as the fuel of your continuous efforts. In our experience, highly driven teams are the ones that do the most and overcome any mishap on the way. Actually, we have worked with leaders who were happy to see bets going wrong because they felt they were closer to the ones that would work well.

The mentality of foxes, the inner power to harness randomness, the ability to manage multiple purposive bets at the same time while embracing creativity, flow and a higher purpose are the core elements of the mindset that contemporary leaders need to employ. This mindset, which promotes positive neuroplasticity and neurogenesis, is remarkably similar to the growth mindset suggested by Dr Carol S Dweck. A mindset greatly affects our behaviour, relationship with success and failure and ultimately our aptitude for happiness and well-being. Dweck (2012) has convincingly explained that there are two main mindsets that people adopt: the fixed one and the growth one. The fixed mindset believes in set abilities, intelligence and personality and does not believe in meaningful change. It results in a life overly dependent on external circumstances rather than on your own actions, which in psychology is also called external locus of control. On the other hand, a growth mindset promotes the freedom of choosing who you are by developing and improving your abilities. Intelligence, personality and capabilities are not static. Failure is just another opportunity to learn and grow while success depends on your own efforts. You can ultimately change and become a better you. This is also called an internal locus of control.

The growth mindset is using neuroplasticity in the best possible way while the fixed mindset is using it in the worst possible way. We have seen again and again that leadership thrives on a growth mindset while it degrades in a fixed one. Which one is your mindset?

Keep in mind

Our brain does not remain biologically unchanged during our professional life. It can do more or it can do less depending on how we use it. Leaders need to take advantage of the brain's neuroplasticity and learn faster, perform better and achieve more. Adopting a higher purpose, getting into the flow, using creativity in all situations, improving their memory and using a diversification, adaptive and growth mindset can help leaders remain always sharp and on top. Your optimized performance will create more devoted followers for you.

References

Advanced Brain Monitoring (2015) Official Website, URL: www.advancedbrainmonitoring.com/, accessed 10 May 2015

American Heritage® Dictionary of the English Language (2011) Fifth Edition, Houghton Mifflin Harcourt Publishing Company, New York

Bechara, A (2011) The somatic marker hypothesis and its neural basis: Using past experiences to forecast the future in decision making, in *Predictions in the Brain: Using our past to generate a future*, ed M Bar, pp 122–133, Oxford University Press, Oxford

Buchanan, DA and Hunczynsky, AA (2010) *Organizational Behaviour*, 7th edn, Prentice Hall, Harlow

Chopra, D and Tanzi, RE (2013) *Super Brain: Unleash the explosive power of your mind*, Rider Books, London

Chrysikou, EG (2014) Your fertile brain at work, *Scientific American Mind*, Special Collector's Edition on Creativity, 23 (1), pp 86–93, Winter

Csikszentmihalyi, M (2008) *Flow: The psychology of optimal experience*, Harper Perennial Modern Classics, London

Dweck, CS (2012) *Mindset: How you can fulfill your potential*, Robinson, London

Entrepreneur (2015) *Creative Genius: Habits and tips from inventive people in the business*, April, pp 56–60

Foer, J (2012) *Moonwalking with Einstein: The art and science of remembering everything*, Penguin Books, London

Fortune (2014) CEO 101: Business Leaders Share their Secrets to Success, 17 November, pp 46–53

Fox, E (2013) *Rainy Brain, Sunny Brain: How to retrain your brain to overcome pessimism and achieve a more positive outlook*, Arrow Books, London

Hamann, S (2001) Cognitive and neural mechanisms of emotional memory, *Trends in Cognitive Sciences*, **5** (9), pp 394–400

Hood, B (2014) *The Domesticated Brain*, Pelican Books, London

Howe, MJA, Davidson, JW and Sloboda, JA (1998) Innate talents: Reality or myth?, *Behavioral and Brain Sciences*, **21** (3), pp 399–407

IBM (2010) IBM 2010 Global CEO study: Creativity selected as most crucial factor for future success, IBM Press News, URL: www-03.ibm.com/press/us/en/pressrelease/31670.wss, accessed 7 May 2015

Kempermann, G, Kuhn, HG and Gage, FH (1997) More hippocampal neurons in adult mice living in an enriched environment, *Nature*, **386** (6624), pp 493–495

Kotler, S (2014) *The Rise of Superman: Decoding the science of ultimate human performance*, Houghton Mifflin Harcourt, New York

Mackay, DG (2014) The engine of memory, *Scientific American Mind*, **25** (3), May/June, pp 30–38

Mourkogiannis, N (2006) *Purpose: The starting point of great companies*, Palgrave/Macmillan, London

Mueller, JS, Melwani, S and Goncalo, J (2012) The bias against creativity: Why people desire yet reject creative ideas, *Psychological Science*, **21** (1), pp 13–17

Pink, D (2009) *Drive: The surprising truth about what motivates us*, Riverhead Books, New York

Scientific American Mind (2014) Upgrading the brain, *Special Issue: The Future You*, **25** (6), November/December, pp 8–9

Seelig, T (2012) *inGenius: A Crash Course on Creativity*, Hay House, Inc, London

Sims, P (2011) *Little Bets: How breakthrough ideas emerge from small discoveries*, Random House, London

Sinek, S (2010) *Start With Why: How great leaders inspire everyone to take action*, Penguin, London

Sinek, S (2014) *Leaders Eat Last: Why some teams pull together and others don't*, Penguin, London

Taleb, NN (2007) *Black Swan: The impact of the highly improbable*, Penguin Books, London

Taleb, NN (2012) *Antifragile*, Allen Lane, London

Taylor, R (2013) *Creativity at Work: Supercharge your brain and make your ideas stick*, Kogan Page, London

The Economist (2010) *The World in Figures: Countries*, The World in 2011, 25-Year Special Edition

SUMMARY OF PILLAR 1: THINKING

TABLE S.1 Summary of pillar 1

Save power for your brain	Higher-level thinking, strong values and immediate feedback are three key aspects for strengthening the leadership willpower muscle. Save power for your brain by: • dealing with ego depletion • dealing with burnout and • dealing with multitasking
Clear your mind	Actively asking questions brings more energy to the rational part of the brain. Four behaviours of asking questions: • the star enquirer • the good listener • the evil interrogator and • the grumpy loner Brain *hijacking* mechanism threatening clear thinking, with important consequences such as erratic behaviour, impaired decision making and performance. *Hijacking* can be managed through a system of distractions and rewards.

TABLE S.1 *Continued*

Focus on performance	Strong purpose can help leaders to see the direction and understand better the decisions made and the decisions that will be made. By getting into the brain state of flow, you and your team can achieve more, faster. Getting into flow requires both internal and external conditions: **Internal** • higher purpose • advanced expertise • low anxiety **External** • new challenge • some control • increased significance Creativity and performance can be enhanced by: – understanding the context of the problem; – avoiding too much analysis; – using the simplest analytical tool; – constantly reviewing the implementation process; – continuously evaluating the results; and – repeating the process as many times as possible. One way to deal with complexity is to employ purposive bets: 1 Place many bets. 2 Minimize the size of the bets. 3 Take the smallest executable step. 4 Calculate affordable loss. 5 Use passion as fuel.

PILLAR 2
Emotions

More emotion, better decisions

04

Always stay cool

He considers himself a great manager. He has held directorial positions in many companies and his track record of results has been stellar, apart from the occasional mishaps, which he has quickly resolved. He has always been very careful in considering options before making a decision, as well as involving the right people. His professional standing and general reputation among his peers is outstanding. He has achieved all this by following a simple rule he learnt when he was young: always stay cool, no matter how critical the situation is.

His motto is that business is only about numbers and that it does not contain emotions. His strong belief is that emotions, both good and bad, can influence the decision-making process and alter what should be a clear and logical outcome, the kind of outcome that markets and shareholders prefer. He has learned to reserve his emotions for the very few moments in life without any risk: a family celebration, a concert of his favourite band, watching sports alone at home. His office is an emotion-free zone and he is very proud of it. He considers his thinking as mathematically oriented, rational and objective, and this has served him well in his career. He uses this principle not only on himself, but for choosing, evaluating and mentoring members of his team as well. Until now, that is.

Two months ago he got a very important call. He was one of the candidates for the top position in the company: the CEO. This was huge for him but, as always, he treated the news with outer calmness. The process of selecting the new CEO was long and challenging and he needed to retain his composure. He was confident that everything that brought him here would make him succeed in the next level as well.

> The company hired an external consultancy agency to run the process of selection. As expected, this included interviews, tests, role-play and meetings with different people within and outside the company. For most of the process he felt that things were going fine. Then the final challenge came. It consisted of a task asking candidates to develop a restructuring plan for a big part of the company. How would they approach it? What would be the main areas of concern and which would be their priorities? It was obvious that this was what the shareholders were considering doing with the company and they want to make sure that the new CEO would tackle it successfully. He started dealing with the task by applying all his knowledge, experience and known cool attitude to every aspect of the project. He identified the market changes that led to the need for restructuring, he evaluated what needed to be done with the current structure, he set new objectives and structures, he estimated budgets, and he developed an effective implementation plan. He even considered negotiations with external partners that have to happen and replacements or transfers of current key personnel. It all looked clean and clear to him.
>
> When the announcement came that he was not the selected candidate he was shocked. Even more so as he was told that he had underperformed in the last task. How could this be? He did everything by the book and it all made business sense. He still cannot fully understand the feedback, especially when it stated that the way that organizations and individuals should understand the role of CEO is not as Chief Executive Officer, but as 'Chief Emotional Officer'. It continued that a top manager, when faced with such monumental challenges as a restructuring programme, should consider first his/her and other people's emotions, attitudes and persuasions in all analyses and plans. Business, it seems, is not only about numbers. It is firstly about people, because people occupy themselves with numbers and make decisions (right or wrong) in businesses. Secondly people do have emotions. A manager should be able to recognize and harness emotions in order to respond accordingly to demanding situations. After this incident, he always tries to consider emotions as part of his core job, both his and those of others. He now understands that there is no situation (professional or personal) not dominated and directed by emotions.

There has been a major misinterpretation of human nature. The advent of science and Renaissance-inspired thinking after the Middle Ages have proclaimed cognitive processes as the pinnacle of human experience, relegating emotions to animalistic states. Bashing emotions out of mental processes ensured objective thinking. 'Stay cool', 'don't be emotional', 'you are over-reacting' and other phrases like these showcase the aggressive and hostile attitude towards emotions from our modern societies and they dominate the traditional jargon of management science since its emergence and theoretical framing by names such as Taylor (1911), Fayol (1930 and 1949) and Weber (1947). The main message was: *The less the emotion, the better the decision.* But can humans live, think and act without emotions? Are emotions evolutionary leftovers that diminish our rational thinking? Should we try to suppress them to become better leaders? Neuroscience answers all these questions with a big NO and the clinical case of a patient called Elliot can help us explain why.

The emotion-run brain

Elliot, in his early thirties, was diagnosed with a fast-growing tumour in the midline area of his brain, which was pushing both frontal lobes upwards. Following successful surgery to remove it, Elliot emerged healthy but changed. Although he was able to think logically and calculate decisions rationally, his actual behaviour was antisocial and far from normal. This change, and similar ones observed in other such cases, led to monumental insights into the role of emotions in decision making and overall in a healthy human life. In a nutshell, it revealed that emotions are at least as important as any other brain function in the way we take decisions and act on them. Decisions without emotions are not just wrong but they may also be dangerous.

Elliot's case, together with the famous Phineas Gage one from the 19th century, who had his skull penetrated by a hot rod while working on a new railroad in Vermont, are described analytically in Antonio Damasio's (2004) seminal *Descartes' Error* book. Damasio, a renowned neuroscientist/neurobiologist, made the following startling statement in his book: 'Reduction in emotion may constitute an equally important source of irrational behaviour.' He convincingly claimed that when emotion is absent there is a counter-intuitive connection with distorted behaviour, which reveals the sophisticated but strong dependence of vigorous reasoning on healthy emotions.

The remarkable insight in Elliot's case, and similar ones from medical history, is that while his high-level cognitive functions were intact after the surgery, his decision making and behaviour were leading to long-term isolation from the people around him. When subjected to tests on perceptual ability, past memory, short-term memory, new learning, language, arithmetic, motor skills Elliot performed exceptionally. He appeared to be a totally normal person ready to perform his daily personal and professional tasks as everybody else. But his actual behaviour was catastrophic for both. How can somebody pass all those tests, appear totally logical but then perform so badly socially and personally? And why is this a phenomenon that neuroscience confirms again and again (Miller et al, 2008)? The answer is as simple and as direct as it can be. The ability to use emotions in decision making and behaviour is as important as, and sometimes even more important than, the ability to use logic. Evolutionarily speaking, emotions developed much earlier in our long history and played a crucial role in making choices. The executive brain, as the latest addition in our skulls, depends greatly on emotions to drive and direct its decisions. Elliot, Phineas Gage and other cases in medical history had actually damaged their neurological paths that inject emotions into their choices and this led to long-term disastrous behaviours. Switching off your emotional brain when taking managerial and business decisions is not only neurologically impossible it is also a very dangerous thing to do.

The biggest impact of emotions in decision making is morality. This is the ability to weight a decision in its later stages based on the possible impact on ourselves and others. It is close to empathy since it helps us share the feeling that an action can cause. People with damaged emotional neural pathways (amygdala-limbic system and hypothalamus-sympathetic response) cannot use their moral compass to guide their decision-making process and behaviour and thus end up making wrong choices both for themselves and others. Actually, there are two medical terms for people who do not have the ability to use empathy and thus take only cold-blooded decisions that eventually work against them and society: psychopaths or sociopaths. Jon Ronson (2011), in his well-known book *The Psychopath Test: A journey through the madness industry* and before him Paul Babiak and Robert Hare (2006) in their famous book *Snakes in Suits: When psychopaths go to work*, argue that reliable characteristics of a psychopath are absence of empathy, remorse and loving kindness. Amazingly enough those are the elements that traditionally schools, MBA programmes and companies tried to push out of future professionals and executives trying to boost their cold-blooded, calculative reasoning. Are we trying to make psychopaths out of managers, leaders and people in general? Modern leaders need to move away from such archaic mindsets and be aware of neuroscience-based realities about what make us great in companies and in life.

Apart from enhancing our decision making, emotions are significant for leadership because they are the bases of motivation. The Latin root of the

English word emotion is 'to move' and this is in plain sight since the word itself includes the term 'motion'. We probably rarely think of it like that but actually the word 'emotion' implies that emotions are the foundation of all our movements. Without emotions we would stay idle, not having the drive to do something, anything. This is why scientists, consultants and entrepreneurs, such as Chip Conley (2012), equate emotions with energy and motion, or as he mentions: 'emotions are vehicles for transforming or moving your life.' Positive or negative emotions spring us into action to either go towards or retrieve ourselves from a situation. Elaine Fox (2013) has elaborated on these two basic brain systems, arguing that they explain much of our human behaviour. Amygdala, as the fear/emergency/avoidance/pessimism centre and nucleus accumbens as the pleasure/excitement/acceptance/optimism centre, drive most of our movements away from or closer to situations. Leaders and managers need to become familiarized with how those systems affect their and other people's behaviour if they are to achieve higher-level decisions.

Emotions run our brain. They are usually classified into three categories. First, the fleeting emotions we feel in any given moment Second, the traits, which are emotions that have long-term presence. Third, the moods that are somewhere between the first two. Neurologically speaking, emotional styles are closer to understanding how our brain wiring affects our emotions and we will start our emotional journey with them.

Emotional style

Psychology has long dealt with emotions but as with everything else it did not have the ability to look into the brain to see what exactly was happening and thus could not provide us with as accurate observations as we have today. The advent of neuroscience and the accompanying technology allowed us to peer into the brain to observe how neurons fire and connect with each other. Research has revealed the neurological basis of emotional predispositions and Richard Davidson, famous for his research on Tibetan monks' brains, has developed a specific categorization of what he calls emotional styles (Davidson and Begley, 2012). An emotional style, in his words, is a:

> ... consistent way of responding to the experiences of our lives. It is governed by specific, identifiable brain circuits and can be measured using objective laboratory methods... Because Emotional Styles are much closer to underlying brain systems than emotional states or traits, they can be considered the atoms of our emotional life – their fundamental building blocks.

Emotional styles mark specific neural pathways that go beyond rudimentary explanations and discussions of emotions so often heard in our everyday lives, within and outside organizations. Based on rigorous and long-term scientific observations Davidson established six emotional styles. Each style lies on a continuum with two extremes on its sides. We believe that these are vital for leaders to understand and direct their behaviour better. Below is an explanation of each style and its relation to leadership, as we have observed it and used it in our practice.

Style 1: Resilience. On one extreme, this style is fast recovery from adversity and, on the other, slow recovery. In essence, this style determines the way we respond to a negative event. If you cannot recover easily from a setback and you keep dwelling on negative emotions then you are closer to the slow recovery extreme. If your ability to function properly goes undeterred and you fight off setbacks quickly then you are closer to the fast recovery extreme. These two extremes depend on the interplay between the amygdala and the prefrontal cortex that we also mentioned in Chapter 2. The more the amygdala is activated over the prefrontal cortex, the more in the slow recovery extreme you will be. The more the prefrontal cortex is activated over the amygdala, the faster you will recover. Bearing in mind the increased dynamism of modern organizations and the small bets approached discussed in Chapter 3, leaders need to move quickly away from disappointments and consider new alternatives with high energy almost daily. Resilience is a key leadership characteristic that allows leaders to find the inner strength to lead themselves and their teams confidently and persuasively to new victories. We recommend to our students and organizational clients to look fast for lessons learned, new alternatives, and reasons to be active as soon as a setback appears.

Style 2: Outlook. This is the popular pessimism–optimism continuum. As with Elaine Fox's dual system of rainy and sunny brain, outlook concerns the way by which we view everyday events and how much are we able to sustain positive feelings or not. Do we tend to search for the negative side of every situation or do we 'always look on the bright side of life'? The interplay of the left prefrontal cortex and the nucleus accumbens, our pleasure centre, is in focus here. The more signals go from the prefrontal cortex to the nucleus accumbens, gearing it towards increased activity, the more you are on the positive extreme. The fewer signals arrive to the pleasure centre, the closer you are to the negative extreme. Our pleasure centre operates mainly under dopamine and opiates, and these two are responsible for different pleasure outcomes. Dopamine activity is associated with anticipation and opiates with pleasure. This means that leaders need to manage those two positive states differently. The feeling of excitement coming from anticipating a result, an event, a meeting, a presentation, is not based on the same chemicals responsible for the feeling of excitement when we actually achieve something. Dopamine is involved in the first and opiates in the second. A positive outlook is associated with higher motivation, energy, creativity, determination and health. We help the people we work for and with to move towards a more positive outlook using opiates, enhancing both anticipatory and achievement emotions. Working towards a presentation in a great teamwork atmosphere and boosting morale by focusing on the upcoming experience and positive result can build dopamine-infused anticipation. Complimenting people's efforts, celebrating successes (even slight ones) and giving appropriate team and individual compensation can build opiate-infused reward (Le Merrer *et al*, 2009).

Style 3: Social intuition. Effectively reading the intentions and emotions of others is essential for great leadership as no one can inspire others while being oblivious to their states of mind. On the one extreme of this continuum is a socially confused person and on the other is a socially intuitive person. When unable to decode someone else's emotions we remain in the dark concerning possible courses of action to amend a situation and boost performance. However, when our sensitivity to other people's emotions is heightened we have the basis of empathy and compassion. Then, we are better fitted to respond appropriately and realistically to a situation. The amygdala is also at play here but this time together with the fusiform gyrus. The fusiform gyrus is located on the temporal and occipital lobes of our cortex and has been associated with various types of recognition. A study published by Josef Parvizi of Stanford School of Medicine and his associates in 2012 proved, through the use of electrodes on a patient's brain, the sole responsibility of this brain region for facial recognition. When the amygdala is activated more than the fusiform gyrus when looking at people's faces, it is a strong indication of a confused style. When the opposite happens, this indicates a highly socially intuitive person. Dr Paul Ekman's (2007) pioneering work on studying human facial expressions all around the world has revealed the evolutionary role of our ability to tell other people's emotional state by observing their faces. Early in our development as a species, when our capacity for language was still limited, we depended on reading people's faces fast and in a subconscious way to adjust our own behaviour and increase our chances of survival. This ability has not disappeared because of our more advanced linguistic skills, but it is also not the same in all people. To improve on this continuum we need to work both on quietening the amygdala, fighting its hijacking potential and trying to become better in reading other people's emotional cues on their face, bodily movements, voice and actions. Observing your friends and close colleagues for emotional cues and discussing with them their emotional state by giving your evaluation, is a good start. However, it has to be done with people who are close and trusted in order to avoid any risks.

Style 4: Self-awareness. Are you able to determine your own emotional state accurately or you are unable to detect or decode your emotions appropriately? Self-awareness is an emotional style with paramount importance to leadership since we can only direct our energy and actions appropriately when we know our thoughts and feelings by detecting accurately the messages our own bodies are sending us. For example, we should understand the impact of a difficult meeting on our emotional state and stop it from spilling over to subsequent unconnected meetings we might have. The main brain region responsible for this awareness is the insula, or insular cortex, which is part of our cerebral cortex. This brain region is related to consciousness and it is increasingly considered as the centre of our sentient self, or the subjective conscious self (see, for example, Craig, 2010 and Gu *et al*, 2013). This is because the insula contains a map of brain organs, receiving and sending signals to them. The higher the activity in the insula, the better the self-awareness, while the lower the activity in the insula, the worse the conscious experience and understanding of our own emotions. The famous ancient phrase 'Know Thyself' inscribed on the Temple of Apollo in Delphi and attributed to Socrates, Plato and Thales among others, is the essence of the self-awareness continuum. At one extreme, our ability for introspection is very weak, severing the connection between our inner self and our conscious self. At the other extreme we can be very sensitive to our internal changes and signals. Our work with managers and leaders in organizations often includes long sessions in discussing levels of awareness of our own emotions in various work situations and their impact on overall behaviour. Taking time to quietly, and in isolation, assess your bodily signals when feeling excited, positively or negatively, can be crucial for understanding your emotional state accurately and acting accordingly. In our experience, self-awareness and purpose are closely related.

Decoding our emotions

Understanding our emotions is crucial for determining our emotional styles. However, this is a very difficult task because of the difference of emotions and feelings. The two are not the same. Emotions are what are happening within us, while feelings are the subjective perception of what is happening in us and its elaboration. According to Damasio (2004), emotions are automated biological reactions in the form of chemical and neural responses to bodily or environmental stimuli. They are generated automatically regardless of whether the stimulus is processed consciously or unconsciously and their role is to make us move/react in order to survive. Feelings, on the other hand, are cognitive representations or perceptions of the changes caused by emotions as well as the evocation of thinking processes and mental states consistent with the emotion. If the insula works well then the awareness of emotions and the generation of corresponding feelings are congruent. We score high on the self-aware dimension. But what happens if those two are not consistent and we misinterpret the emotion for something that it is not, regardless of the working of the insula? Can the perception of our emotions be tricked by environmental and other conditions? And if yes, what can we do about it?

In a 1974 publication, psychologists Dutton and Aaron described their now legendary Capilano Suspension Bridge Experiment. In the experiment, the researchers exposed a group of young men to an attractive female researcher conducting a survey and wanted to measure how many of them called her back at the end of the experiment. The female researcher gave the male participants her phone number as part of the process to answer any additional questions after the experiment had taken place. One group of young men met the female researcher in the middle of an unstable suspension bridge and the other on stable ground. The results were clear: those who were on the suspension bridge contacted the female researcher more than those who were on safe ground. They were fooled by their own perception. The group on the dangerous bridge misattributed their feeling of arousal/excitement to the sexual attractiveness of the female researcher, rather than to fear of survival. The issue is that the physical sensations of the two within our bodies are similar and an external stimulus nudged them towards the wrong direction. A great and graphic case of inconsistent feelings and emotions.

Our recommendations for improving the perception of your own emotions, especially in professional situations, is called *3D+3L Approach* and includes three types of descriptions and three types of listening:

1D *Description of the situation.* Take time and describe in detail a situation that made you experience intense feelings. Pay attention to causes, other people's roles and the physical surroundings. What forced you to feel like that? Consider this to be the equivalent of strategic environmental analysis. This description needs to happen relatively close to the event.

2D *Description of reactions.* Continue with the detailed account of the behavioural response that the feeling produced, first in you and then in the others around you (if applicable). Track, in a step by step way, everything that followed from having the feeling to playing it out. Write down both seemingly unimportant actions that only you know about and significant ones that involved others. Why did you behave that way? This also needs to be done close to the event.

3D *Description of correlations.* Create a list of events that had similar characteristics or similar behavioural outcomes. Are there any connections? Also, try to link your chains of thought to your feelings, the situation and the behaviours that followed. Are there any patterns? This is a more introspective analysis that is usually ongoing and generally takes more time than the previous ones.

1L *Internal listening.* Look inside yourself with minimum interruption and with genuine interest to help you establish a stronger link between emotions and feelings. Silencing our typical chatter and ensnaring our sensitivity to raw emotions before we jump into easy conclusions are necessary steps in improving our understanding of our emotions. Mindfulness has become a global trend for CEOs and executives because of its promise of increasing self-awareness, as explained by a pioneer in the field, Maria Gonzalez (2012). Give yourself the chances, planned and unplanned, to listen to your inner self.

2L *External listening.* Listen to others carefully and try to understand their viewpoint, without letting your strong feelings interfere/misinterpret the conversation. This is key to having a clear emotional picture of the situation. Furthermore, asking others to evaluate and give you feedback on your emotional state can help bypass our internal miscommunication of our own feelings. Don't hesitate to ask the ones closer to you.

3L *Constant listening.* Developing a mental radar that constantly scans internally and externally for emotional cues and their impact on you and those around you is the ultimate goal for self-awareness. Your own personal 'emotions observatory' should be established in your mind that will relentlessly evaluate important and less important moments for emotional reactions and that will set a knowledge directory for emotional patterns and related behaviours.

Action box: the 3D+3L approach in practice

Think of a recent emotionally challenging situation that you have passed through with more or less success. Then, use the three types of descriptions and three types of listening above in order to write down how this situation could be interpreted. What are your conclusions? Are there things that you could do differently regarding this situation?

Style 5: Context sensitivity. Are you emotionally harmonized with what is happening around you and are you behaving accordingly? Or are you usually caught unaware of the situation, adopting an inappropriate behaviour that sometimes makes you embarrassed? On one extreme in this continuum is people tuned in to the environment and, in the other, tuned out. Tuned in means emotional synchronization while tuned out means emotional disconnection with a situation. We have seen, for example, managers making jokes in a crucial moment of a presentation, or being over-serious when a business partner tried rightfully to lighten up a conversation. The brain region associated with such emotional synchronization is the hippocampus. We have already discussed the hippocampus in relation to memory formation, but it plays also a vital role in fitting or tuning behaviours to contexts or situations. A weaker, or even smaller, hippocampus can mean completely inappropriate emotions, and subsequent behaviours, to external conditions. Furthermore, contextual learning is diminished in such cases. Thus we need to make sure that our internal state and external requirements fit both in order to behave accordingly and to take away the right lessons from the situation. Some of the recommendations made in the 'Decoding our emotions' box apply here too. Listening to others and creating as accurate as possible descriptions of what happened can lead us to pattern recognition, learning and appropriate behavioural adaptations in the future. A note of caution: if you are not sure if what you sense in the meeting room is tension or positive anticipation then better to wait for more cues until you make a decisive move. If you feel lost constantly, then focus on working on moving up the continuum and towards a more *tuned in* style. But be careful, too much sensitivity can increase stress responses since it is associated with anxiety and feelings of inadequacy. As Ellis and Boyce (2008) have argued, although sensitivity to context in our brains is evolutionary and invaluable for learning to deal effectively with the environments we find ourselves in, it follows a u-curve in relation to stress, meaning that low sensitivity produces high stress as does high sensitivity. Being a little disconnected, or not fully connected, is a great asset for modern leaders in our unpredictable and stressful modern business environment. How disconnected is something that each leader needs to determine individually based on his/her experience and personality as well as on the situation.

Style 6: Attention. On the one extreme of this dimension is a focused person while on the other extreme is an unfocused one. Do you have trouble focusing on the person talking to you from across the meeting table? Do you feel your mind wandering every time you look at the detailed monthly results or the latest employee engagement scores from HR's research? Are you often annoyed at your open plan office because the noise does not allow you to do your job properly? Then you are closer to the unfocused end. In such cases your emotional awareness will be limited and your behavioural response unsuitable to the situation at hand. The prefrontal cortex, or executive brain, is responsible for focus since higher activity can produce a condition called 'phase-locking', which is the ultimate focus. In this condition the engagement of the prefrontal cortex is perfectly timed with the external stimulus, meaning that you are mentally in line with what you are paying attention to. This kind of 'phase locking' is much admired in corporations. A Bloomberg article authored by Buhayar (2013) reported on the popularity in CEOs' speeches of the phrase *laser focus*. Being laser-focused on execution, growth, new products and market opportunities among others has apparently entered the jargon of top-level communications making sure that distractions will not be tolerated. However, too much focus is not always an advantage. A study by neuroscientist Jason Gallate and his associates (2012) revealed that creative people benefit the most when taking a break from focusing on a given problem and that this break allows for non-conscious processes to take over and provide insightful solutions. Too much focus inhibits creativity, but loose focus inhibits performance and productivity. At the same time, there are situations when extreme focus is needed, such as in certain negotiations, and thus the intensity of focus has to be determined on a case-by-case basis.

Continuous practice is the key towards moving to the one or the other extreme in all of the styles above, making use at the same time of something we referred to in Chapter 3 as neuroplasticity: the ability to change our neural pathways and thus change positions in the emotional styles' continuums. Although it might be challenging to go from one extreme to the other, moving along the line is both possible and advisable. We have seen it happening both in ourselves and in the people we have been working with. Putting this as a regular practice in your agenda will provide more successful results.

Action box: measure emotional styles in teams

Go through all six dimensions by Davidson and score yourself from 1 to 5, 1 being the low extreme and 5 being the high extreme. Reflect on every score. Write down a few bullet points on the scores you want to retain and the ones you want to change, upwards or downwards. Ask your closest associates to score you on all six and observe any deviations between your scores and those your associates gave you. Why do you think they exist if they do? You can do this exercise as a team as well, each one first scoring themselves and then scoring each other. An open discussion based on trust and positive feedback can do a lot to help each other improve on the dimensions. This is not to be initially tried in the team setting without first having professional and experienced help.

From mood to great feelings

The long-term effects of emotional styles on how we think and behave in our organizations, although not the same, resemble the long-term effects of personality: they follow us through a large part of our lives. They can change to better fit our leadership challenges but this takes time and much targeted effort. In a more short-term horizon, our emotions are highly influenced by our moods and modern leaders need to be aware and try to control them since they form the fertile ground for our everyday emotional reactions. Although themselves hidden, they are responsible for most of the emotions we experience in our workplace. According to one of the leading experts on moods, Dr Liz Miller (2009):

> We all have moods all the time... Moods are an internal measure of how we are. We do not express our moods directly. Instead we express them indirectly in the way we think, communicate, behave and see the world.... Almost all anger reflects an underlying irritable, anxious mood. This mood has provided the soil that allowed the emotion of anger to grow.

It is your general mood that will determine greatly the type of emotions you will express in any given moment. Thus, by getting into the right mood and avoiding counterproductive ones, you can reach for the right set of emotions in any situation and use them to further promote your leadership abilities. For example, how much better can you manage your emotions and valuable contribution in a difficult meeting about the latest increase in employee absenteeism if you are in a positive mood compared with a negative one? We have encountered countless managers who have confided in us that they wished they had reacted differently in a stressful situation, but that the overall bad mood in the company 'during those days' did not allow them to do so. If they knew how to set their mood to positive they would have dealt much better with any challenging situation. This is because, in the long list of benefits, nurturing a positive mood is integral to concentration and clear thinking, managing your behaviour, creating meaningful and lasting relations, flexibility and perseverance in pursuing goals and work satisfaction. In a particularly interesting study, Ruby Nadler of the Western University Canada and her colleagues (2010) have demonstrated that people in a positive mood outperform those in neutral or negative moods. Those in positive moods show higher cognitive flexibility and do much better in tasks that require heavy use of the prefrontal cortex, our executive brain, such as hypothesis testing and rule selection. This has a profound effect on management since a positive mood emerges as a prerequisite not only for the softer aspect of the job, such as keeping the team upbeat and motivating individuals, but mainly for better utilizing our cognitive processing capacities. That is, to think better and come up with better solutions to difficult problems. Good mood means good thinking.

Miller's (2009) now famous categorization of moods uses two variables to plot a 'mood map': energy on the one axis and well-being on the other. The first axis (vertical) has high energy on the top and low energy on the bottom. The second (horizontal) has negative well-being on the left and positive on the right. Energy relates to the neurotransmitter dopamine, which is about arousal/excitement and well-being to the system Serotonin-Endorphin which is about positive feeling, good functioning of internal organs and pain relieving. The four quadrants of the mood map and the feelings belonging in each one are explained below, following our practice with managers worldwide:

Q1 The rocket mood: high energy, positive well-being quadrant. Strongly motivated, highly pleased and continuously excited leaders that drive everyone forward with passion and conviction. Fast-moving, with unlimited energy and a great smile on their face, leaders in the rocket mood are the guiding and unstoppable force of their teams. This mood rarely gets you tired and when it does you bounce back with even more resourceful ideas and precious emotional support for others.

Q2 The guru mood: low energy, positive well-being quadrant. Calm, highly content and satisfied leaders who create the conditions for thoughtful work, a peaceful office environment and the unbeatable feeling of safety. Leaders in the guru mood are an effective weapon against the craziness of a situation and create the necessary stability for reaching wise decisions with high long-term effects.

Q3 The downer mood: low energy, negative well-being quadrant. Physically and mentally exhausted, often sad and even depressed leaders that cannot hide their boredom alongside their negative feelings at work. Leaders in the downer mood negatively affect those around them, sucking energy and diffusing pessimism.

Q4 The panic mood: high energy, negative well-being quadrant. Tension, irritation, fear and frustration are a few of the powerful but negative feelings expressed by leaders in this mood. A jumpy attitude, a strong focus on self-preservation and the image of running like a headless chicken are associated with leaders in the panic mood that seem to make everyone around them nervous and insecure.

A leader does not have to be at these extremes though to identify his/her mood as the *Rocket*, the *Guru*, the *Downer* or the *Panic*. Even milder versions capture the essence of each category. Keeping yourself on the positive side and increasingly on the high energy side is of paramount importance since even solving difficult math problems and speaking in public are associated with appraising your feelings as excitement instead of anxiety. Alison Wood Brooks (2013) from Harvard Business School found that when there is emotional arousal because of an upcoming important challenge, such as a presentation or a test, people do much better when they interpret this as excitement than anxiety. And a positive, Rocket mood helps you do exactly that. She also found that people trying to use the old strategy of calming down, do badly. Another nail on the coffin of the 'staying cool' approach.

Action box: check your mood

Mentally revisit the last two weeks at work. Did you have a prevailing mood throughout or did you experience different moods at different times? Write down the mood(s) you recognize and try to categorize it/them using the model explained in this part of the book By using the four quadrants of the mood map try to describe your experience based on Q1, Q2, Q3 or Q4. In order to confirm your conclusions it is always a good strategy to ask the people that you trust and interact with more, especially those close to you in the period of time that you try to evaluate (the last two weeks). Write down the causes and outcomes. Repeat the exercise once every month. After a few months try to identify patterns based on the following questions:

- Are specific moods appearing because of specific causes, having specific outcomes?
- Why do you think this is the case?

Since the link between positive moods and better performance is scientifically established, your goal is to move towards sustained positive moods such as the Rocket and the Guru. Achieving a positive mood requires changes in the environment, in awareness/knowledge, in relationships and in emotional styles as described in the previous section. Do not allow your mood to damage your leadership potential. On the contrary, put your mood to work for you and your emotions. Proactively, and after you become comfortable in detecting your mood, try to do so before an important meeting, presentation, event and go in with the most appropriate one.

Moods are not as fixed as emotional styles and personalities, but they can set themselves in for a long time if not challenged. Having a constant negative mood is a sign for the need for change and leaders should often consider their mood to make sure they give themselves the chance to experience the right emotion at the right time.

EQ as an empowerment qualification

Since the mid-90s when Daniel Goleman (1995) spread the word for the merits of emotional intelligence, or emotional quotient (EQ), with his international best-selling book, there have been countless other books, white papers, academic research papers, consulting and training events, services and experts on the topic. Multinational corporations have adopted the practice of evaluating their executives on emotional intelligence and the concept has even been found to help managers fulfil their leadership potential since, as Dijk and Freedman (2007) illustrated, employees become more skilful in emotional literacy and subsequent thinking as they climb up the organizational ladder. For us, EQ was – and still is – seminal in changing the out-dated mindset in societies and businesses that emotions are inappropriate and even destructive. As excellently put by Cooper and Sawaf (1996) in their early and highly influential book *Executive EQ*, we need to move on from considering emotions as:

- an interference with good judgement to essential to good judgement;
- a force of distraction to a force of motivation;
- a sign of vulnerability to a sign of vitality and presence;
- an obstruction to reasoning to an enhancement and acceleration to reasoning;
- a barrier to control to the foundation of trust and connection;
- an inhibition to objective data to a significant information and feedback mechanism;
- a complication to planning to the main spark of creativity and innovation;
- as weakening our attitudes to activating our morality;
- as undermining our authority to boosting our influence without authority.

We have to attribute to the EQ movement the wave of change in organizations worldwide of accepting emotions as a reality that needs to be incorporated into both mindsets and processes. Boyatzis and Goleman (1996), in their book titled *Emotional Competency Inventory*, suggest an emotional intelligence model for leaders developed as the first holistic approach in managing emotions that inspired the masses and made its way so profoundly into corporations, big and small. Both scholars and their associates expanded and developed this approach further in later works (Goleman *et al*, 2002; Boyatzis and McKee, 2005). This holistic emotional intelligence model, which we also continue practising in consulting, coaching and training engagements consists of four major clusters, two of which concern our internal world and two the external:

Cluster 1: Self-awareness (internal awareness) is about being able to recognize our own emotions and consists of:

- emotional self-awareness: recognizing our emotions and their effects;
- accurate self-assessment: knowing one's strengths and limits; and
- self-confidence: a strong sense of one's self worth and capabilities.

Cluster 2: Self-management (internal management) is about being able to change our own emotions and consists of:

- adaptability: flexibility in dealing with changing situations or obstacles;
- self-control: inhibiting emotions in service of group or organizational norms;
- optimism: a positive view of life;
- initiative: proactive, bias toward action;
- achievement orientation: striving to do better;
- trustworthiness: integrity or consistency with one's values, emotions, and behaviour.

Cluster 3: Social competence (external awareness) is about being able to recognize emotions in others and consists of:

- empathy: understanding others and taking active interest in their concerns;
- service orientation: recognizing and meeting customers' needs;
- organizational awareness: perceives political relationships within the organization.

Cluster 4: Relationship management (external management) is about being able to change emotions in others and consists of:

- inspirational leadership: inspiring and guiding groups and people;
- developing others: helping others improve performance;
- change catalyst: initiating or managing change;
- conflict management: resolving disagreements;
- influence: getting others to agree with you;
- teamwork: creating a shared vision and synergy in teamwork, collaborating with others, building relationships and networks.

In a nutshell, EQ helps leaders become aware of and manage their own emotions and the emotions of others. It starts from inside you and spreads out towards the people around you. Empathy, active listening and intrinsic motivation are only a few of the keywords that EQ spearheaded bringing into our lives and they seem to be key elements in leadership rhetoric. As Antonacopoulou and Gabriel (2001) argue, these elements are necessary in leading with care. They also suggest that a leader needs to have guts in order to lead. Having the guts to lead means emphasizing the following three emotional dimensions:

1 Emotions as coping mechanisms (emotions can help us to adapt to changing circumstances).
2 Emotions as transitional qualities (emotions preserve what a person values in different circumstances).
3 Emotion as a system of reactions (emotions support interpretations of situations).

In other words, emotions are key elements in engaging leadership. In our experience, EQ is a true empowerment qualification since without EQ a leader is unable to navigate through and appropriately manage emotional styles and moods internally and externally. Any capacity to inspire, influence and move others is minimized and the person will soon feel outworked, outperformed and eventually like an outcast. On the contrary, EQ brings an ideal predisposition for working with emotions and a high probability for succeeding in the long run.

Regardless of the widespread appeal of EQ and its two decades of influence we still have to experience managerial and organizational systems that take into account fully the teachings and practices of EQ. There are training and other initiatives that have moved towards this direction but overall organizations still look and feel emotions-averse more than emotions-run. Since our brain uses emotions in such diverse ways and boosts through them both thinking and performance we hope that a new wave of emotional intelligence will hit companies. This new wave is the neuroscience one. All in all, we cannot be overemotional about the importance of emotions in leadership.

Keep in mind

A brain without emotions is a malfunctioning brain and this is something that organizations, leaders and managers have to understand quickly in order to calibrate better towards a brain-based leadership approach. Emotions help us think faster and with morality, move ourselves and others and boost our cognitive capacity for advanced performance. Being aware of your emotional styles and moods, and applying a holistic approach that takes into account both the internal and external worlds, can help you bypass the out-dated notion of incompatibility of emotions and business. Leadership is primarily an emotional skill and CEO truly means Chief Emotional Officer in any organizational level.

Boost your brain: identify emotions in leadership

Think of a leader for whom or with whom you have worked a lot and especially someone that you would gladly work with or for again. Then, think of a person in a leadership position that you either try to avoid, or left little impact on you. Then try to compare both people by writing down how each of them acted in their daily professional lives. Also, think and write down how they related to others. What lessons can be learnt from this comparison? Can you identify the power of emotions? Consider your conclusions and use them next time.

References

Antonacopoulou, EP and Gabriel, Y (2001) Emotion, learning and organisational change: Towards an integration of psychoanalytic and other perspectives, *Journal of Organisational Change Management*, **14** (5), pp 435–451

Babiak, P and Hare, DR (2006) *Snakes in Suits: When psychopaths go to work*, HarperBusiness, London

Boyatzis, RE and Goleman, D (1996) *Emotional Competency Inventory*, The Hay Group, Boston, MA

Boyatzis, R and McKee, A (2005) *Resonant Leadership: Sustaining yourself and connecting to others through mindfulness, hope, and compassion*, Harvard Business School Press, Boston

Brooks, AW (2013) Get excited: Reappraising pre-performance anxiety as excitement, *Journal of Experimental Psychology: General*, 143 (3), pp 1144–1158

Buhayar, N (2013) 'Laser-focused' CEOs proliferate as jargon infects speech, *Bloomberg online*, URL: www.bloomberg.com/news/articles/2013-09-11/laser-focused-ceos-multiply-with-promises-from-ipads-to-macaroni, accessed 4 June 2015

Conley, C (2012) *Emotional Equations: Simple truths for creating happiness+success*, Simon and Schuster, New York

Cooper, RK and Sawaf, A (1996) *Executive EQ: Emotional intelligence in leadership and organizations*, Perigee, New York

Craig, AD (2010) The sentient self, *Brain Structure and Function*, 214 (5–6), pp 563–577

Damasio, A (2004) Emotions and feelings: A neurobiological perspective, in *Feelings and Emotions: The Amsterdam symposium*, eds AS Manstead, N Frijda and A Fischer, pp 49–57, Cambridge University Press, Cambridge

Damasio, A (2008) *Descartes' Error: Emotion, reason and the human brain*, New Edition, Penguin, New York

Davidson, JR and Begley, S (2012) *The Emotional Life of Your Brain: How to change the way you think, feel and live*, Hudson Street Press, New York

Dijk, CFV and Freedman, J (2007) Differentiating emotional intelligence in leadership, *Journal of Leadership Studies*, 1 (2), pp 8–20

Dutton, DG and Aaron, AP (1974) Some evidence for heightened sexual attraction under conditions of high anxiety, *Journal of Personality and Social Psychology*, 30, pp 510–517

Ekman, P (2007) *Emotions Revealed: Recognizing faces and feelings to improve communication and emotional life*, Owl Books, New York

Ellis, BJ and Boyce, WT (2008) Biological sensitivity to context, *Current Directions in Psychological Science*, 17 (3), pp 183–187

Fayol, H (1930) *Industrial and General Administration*, Sir Isaac Pitman & Sons, London

Fayol, H (1949) *General and Industrial Management*, Sir Isaac Pitman & Sons, London

Fox, E (2013) *Rainy Brain, Sunny Brain: How to retrain your brain to overcome pessimism and achieve a more positive outlook*, Arrow Books, London

Gallate, J, Wong, C, Ellwood, S, Roring, RW and Snyder, A (2012) Creative people use nonconscious processes to their advantage, *Creativity Research Journal*, 24 (2–3), pp 146–151

Goleman, D (1995) *Emotional Intelligence: Why it can matter more than IQ*, Bantam, New York

Goleman, D, Boyatzis, R and McKee, A (2002) *Primal Leadership: Realizing the power of emotional intelligence*, Harvard Business School Press, Boston

Gonzalez, M (2012) *Mindful Leadership: The 9 ways to self-awareness, transforming yourself and inspiring others*, John Wiley and Sons

Gu, X, Hof, PR, Friston, KJ and Fan, J (2013) Anterior insular cortex and emotional awareness, *Journal of Comparative Neurology*, **521** (15), pp 3371–3388

Le Merrer, J, Becker, AJJ, Befort, K and Kieffer, LB (2009) Reward processing by the opioid system in the brain, *Physiological Reviews*, **89** (4), 1379–1412

Miller, L (2009) *Mood Mapping: Plot your way to emotional health and happiness*, Rodale, London

Miller, M, Bentsen, T, Clendenning, DD, Harris, S, Speert, D and Binder, C (2008) *Brain Facts: A primary on the brain and nervous system*, Society for Neuroscience, Washington

Nadler, RT, Rabi, R and Minda, JP (2010) Better mood and better performance: Learning rule-described categories is enhanced by positive mood, *Psychological Science*, **21** (12), pp 1770–1776

Parvizi, J, Jacques, C, Foster, BL, Withoft, N, Rangarajan, V, Weiner, KS, Grill-Spector, K (2012) Electrical stimulation of human fusiform face-selective regions distorts face perception, *Journal of Neuroscience*, **32** (43), pp 14915–14920

Ronson, J (2011) *The Psychopath Test: A journey through the madness industry*, Picador, London

Taylor, WF (1911) *The Principles of Scientific Management*, Harper & Brothers, New York and London

Weber, M (1947) *The Theory of Social and Economic Organization*, Free Press, New York

Right emotion, right action

05

Fear equals hard work

No one should underestimate the power of fear in the workplace. Especially the executive of our real-life case who, during her early professional years, worked closely with a boss who led with an iron fist and always did her best to spread fear and terror throughout the office. Although disagreeing with such methods in principle, she instinctively learned from this boss that people will focus more, work harder, and stay longer out of fear rather than incentives. She just had to observe her own reaction to this emotional environment created by her boss to understand how effective fear was. Sure enough, some people could not handle it and either broke down or left (or both). But the necessity for such an approach was apparent by the challenges the company faced at the time: the fierce competition, political turmoil and high stakes were just too much for a fluffy, soft approach. This was a game for strong people and only the strongest would survive. And so she did.

Becoming a top executive in the same company took her several years. But years did not take away the main lessons she learnt in those early days. Fear is her biggest weapon, both for her own motivation and for her subordinates. A generally fearful culture in her department helps her to keep people committed, productive and more easily managed. Fear keeps her in control, and control is all she needs to manage herself and others. She believes in a simple equation: fear leads to control and control leads to effective management and effective management leads to great performance. However, she is not completely blind to the advantages of positive motivators. She uses both informal and formal positive measures when needed to achieve particular objectives, such as attracting new personnel, keeping her team from constant emotional collapse, preventing people from leaving the company and sometimes for feeling more humane herself. After all, she does believe that humans are inherently good and kind and she is a great person to most of her friends and family. It is just that serious company work requires a 'serious' approach. Thus, the stick is for keeping people in order and constantly running, and the carrot for attracting new ones and stopping the good ones leaving. The 'carrot and stick', or better, the 'stick and the carrot' are her leadership toolbox. So simple, so effective.

However, she started facing problems when a new chief of human resources (CHR) pointed out in a high-profile meeting that they could not afford any more great talent fleeing the company. This was because, according to the latest corporate culture survey, the climate in many departments had to urgently change to become more positive. The department that scored worse in this internal research was hers. Also, her department had the highest rate of talent turnover. Following those numbers, the CEO requested a personal meeting with her. In that meeting she had the opportunity to explain her approach and how effective it had been up until now, and that she saw no reason for changing her 'stick and carrot' approach. She claimed she knew how to use emotions to motivate her employees as well as herself and that she was upset about the whole new situation. The CEO, in cooperation with the new CHR, asked her to go through intensive emotions training and assigned her a new professional coach, specializing in emotional management. After some time, she claims her life has changed and that she now sees how beautifully complicated human emotions are. She is a much better manager and a great leader to her team and to her colleagues. It is clear to her now that serious business needs a serious approach to emotions and not the simplistic one she was applying.

Why have management and business in general been allergic to a serious and deep discussion of emotions for so long? And why, when acknowledging that emotions might actually play a role, do they use humiliating concepts such as the carrot and the stick, which are more appropriate for donkeys than for real people? Fear, as a core management tool, is widespread in organizations of any type and size as our experience around the world suggests. Simplistic views of emotions in companies are dominating management styles, partially because the neurobiological basis of emotions is missing from basic, and even from advanced in many cases, management education and training. In other words, the majority of management education programmes around the world, and especially MBAs, are based on principles and practices supported by the traditional jargon of management science neglecting critical findings of other disciplines that specifically call for emotional and social intelligence to become integral skills in management and leadership (Heiss, 2014). What emotions do we have and how can leaders acknowledge them and use them within themselves and others in their organizations? Emotions induce tremendous moving power and modern leaders can no longer be oblivious of how they actually work and of the combinations that can work better for them in their organizations.

The basic emotions in the brain

There is a wider consensus, as seen in the previous chapters, that emotions are necessary for human motivation and action. Without emotions there is no urgency to behave in any way; to transfer from a stable condition of homeostasis ('standing still') to a more dynamic condition of movement. So, the higher the need for change, even for constant transformation as in many contemporary industries, the more urgent the need for emotions to keep people changing, taking the initiative and bringing new solutions to new problems. The brain is actually built for that, but are we using it appropriately?

The emotional state in the brain is often described as a continuum, with positive and negative on the two extremes and neutral in the middle. Both extremes are there to make us move, either towards or away from an object, a person or a situation. Ellaine Fox (2013) describes these two conditions as the avoid/approach systems in our brains which, depending on the decoding of the information reaching the brain, activate different circuits to make us act accordingly. The important point here though is that the number of emotions that exist on this continuum and form the toolbox for leaders to use on themselves and others to activate avoid/approach systems, is established in psychology and neuroscience. We will examine a few of them here, highlighting both similarities and differences in order to demonstrate how they work.

We have previously separated emotional states (emotions) from their subjective perception (feelings), their midterm patterns (moods) and their long-term, personality-like models (emotional styles) and we analysed moods and styles in Chapter 4. Dealing now with the key element, emotions themselves, there is an obvious effort in science to identify what the core emotions are that drive our behaviour. This is because, although we use various names to describe our internal states in our everyday lives, scientists discovered that there is actually only a small number that dominate our brains and bodies. Knowing those and how they relate to each other can help us deal more effectively with everyday personal and business challenges. Both Ekman and his teacher Tomkins studied facial expressions to reveal basic human emotions. Ekman's (2007) model of universal facial expressions found that, all around the world, six basic emotions are mentioned – anger, disgust, fear, happiness, sadness and surprise – while Tomkins' (2008) model, originally published in 1962 and 1963, identifies nine basic emotions or 'affects' as he calls them, some in low/high pairs, with each pair having a low/high expression, which scholar Nathanson calls: 'hard-wired, pre-programmed, genetically transmitted mechanisms that exist in each of us' (Nathanson, 1992). These are:

1 *Enjoyment (low)/joy (high)*. This is a positive reaction to success and the higher the emotion the stronger the willingness to share.
2 *Interest (low)/excitement (high)*. This is a positive reaction to a new situation and the higher the emotion the stronger the participation and engagement.
3 *Surprise (low)/startle (high)*. This is a neutral reaction to sudden change that can reset our impulses.
4 *Anger (low)/rage (high)*. This is a negative reaction to threat and the higher the emotion the stronger the physical or/and verbal attack.
5 *Distress (low)/anguish (high)*. This is a negative reaction to loss leading to mourning.
6 *Fear (low)/terror (high)*. This is a negative reaction to danger and the higher the emotion the higher the impulse to run or hide.
7 *Shame (low)/humiliation (high)*. This is a negative reaction to failure creating the need to review one's own behaviour.
8 *Disgust*. This is a negative reaction to a bad offering, not necessarily restricted to food, and it motivates to expel and reject.
9 *Dissmell*. This is a negative reaction to a repelling situation and it strengthens the impulse to avoid and keep things/people at a distance.

Both Ekman's six and Tomkins' nine basic emotions have influenced both scientific and non-scientific professionals around the world, mainly because of their straightforwardness and direct relation to the human face. Those sets of basic emotions involve both macro and micro-expressions of the face, the latter not able to be consciously detected or manipulated, and thus reveal deeper brain structures and how they impulsively react to stimuli. The latest research on Ekman's model reduces the number of basic emotions on facial expressions down to just four: happiness, sadness, fear/surprise, anger/disgust (Jack et al, 2014). Utilizing the latest technology and methods, this study demonstrated that happiness and sadness show unique facial signals over time, while fear and anger share initial signals with surprise and disgust respectively. Although, as the cycle of an expression matures, those pairs separate again and it is evident that facial signals are designed by both biological and social evolutionary pressures to optimize their function as expected by evolutionary predictions. So, fear and surprise start the same, as do anger and disgust, probably highlighting that when it comes to dealing with a negative stimulus imminently compromising our survival, fast reaction and quick signalling is of elevated importance.

The basic emotions and the relations between them are intrinsically linked to chemicals in the brain. Hugo Lövheim's (2012) model is the first one to comprehensively link comparative levels of the three monoamine neurotransmitting chemicals in our brain: serotonin, dopamine and noradrenaline. Lövheim combined Tomkins' nine basic emotions with all three neurotransmitters and showed which chemicals have to be high and which low for every emotion. Serotonin, for example, a chemical related to being able to control our behaviour, think clearly, regulate our mood and especially avoid aggression (Badawy, 2003) is high in the positive emotions of joy and excitement, as well as in surprise and disgust. This is probably because in such emotional states we have the least need to show immediate aggression and the most to show openness and good decision making. On the other hand, dopamine, the arousal hormone that gets us ready for action through anticipating results, is high in the two positive emotions of joy and excitement but also in the negative emotional pairs of fear/terror and anger/rage. This shows that dopamine is not just for positive (reward-related) situations but also for negative ones that need our immediate prediction of possible outcomes and appropriate response (Schultz, 2002). Finally, the stress hormone, noradrenaline (also known as norepinephrine), is high on distress, anger, interest and surprise, an interesting mix of negative, positive and neutral emotions. This is because noradrenaline is highest present when the situation we face is drastically different from the one our memory and expectations prepared us for, thus we go through 'unexpected uncertainty' (Yu and Dayan, 2005). So surprise, interest, anger and distress are naturally linked with high noradrenaline. In such cases we need to re-examine what we know and maybe learn new methods and behaviours. Lövheim's model helps us look deeper into the neurobiology of emotions and understand the brain changes on a chemical level occurring when we experience them.

Our premise is that regardless of the model followed, leaders and managers need to understand the basic emotions and separate them from feelings, moods and styles in order to be able to work with them more effectively in various challenges. Being aware of the automatic reactions emotions have on behaviour, leaders can better steer their actions and the actions of others. The oversimplified and sometimes offensive model of the carrot and stick as a caricature representation of the avoid/approach systems in our brain has to be urgently replaced by a more realistic and complex knowledge of emotions and how they actually affect our motivation.

Applying Tomkins' model in organizations

We have been applying Tomkins' model of nine core emotions for some time now with our clients and the following key insights on each one are readily available for implementation by any manager and aspiring leader. We have remodelled it to fit business and institutional environments.

Dealing with *positive emotions*

1 *Celebrate*. With any success, even the smallest, organize a celebration both for yourself and your team. Although the size and type of celebration should be directly related to the size and type of success, never allow a success, regardless of how small, pass without celebrating it. Share the joy and your team will share it inside and outside the organization with multiple positive effects. Celebration is a social process not an individual one.

2 *Explore*. In accordance with the growth mindset described in the previous chapter, new data, people, situations etc should be looked at with genuine interest and even with excitement. According to the Tomkins Institute (2014) such an approach directly connects effective thinking with a good emotion since this is the process by which learning is rewarding. Improve your thought and memory by excitedly exploring the world around you, constantly. This is the safest way to enhanced engagement. Furthermore, Jaak Panksepp, the world's authority in affective neuroscience, has indicated that in his model of seven primal emotions, namely seeking, rage, fear, lust, care, panic/grief, and play, seeking is probably the strongest one since it is the seeking circuits firing in the brain when we get excited from new intellectual

connections, novel ideas, cutting-edge technologies, and when we enthusiastically search for meaning (Panksepp and Biven, 2012). That is, seeking keeps us motivated in the direction of what we are seeking.

Dealing with the *neutral emotion*

3 *Stop-and-think*. When experiencing sudden, brief, unexpected changes it is advisable to refrain from a knee-jerk response and consider the situation closely before action. Especially when looking momentarily inside you fails to produce convincing insights fast (based on memory and experience), stopping and giving time to yourself and your team to think and better consider the situation is the best strategy. Importantly enough, acknowledging this fact to ourselves and the team does not make us weaker leaders but wiser ones.

Dealing with *negative emotions*

4 *Recharge*. Businesses and organizations of every type face constant changes and increasing threats from all around. Both geographically and industrially speaking, new competitors arise from anywhere and threaten the very existence of even established corporations. Threats also exist within companies, with people trying to develop their careers at the expense of others. The brain reacts with anger when there are too many such threats and/or a few big ones. It stems from too much firing of neurons in the brain and the resulting inability to resolve the challenge at hand effectively. So, when feeling rage and wanting to attack in full force, just ask yourself whether this is actually all about an overburdened brain and if attack is really the best way forward. Wanting to attack reveals that the one under attack is actually you and so you need to reorganize yourself and the team to make sure your response will be optimized. It also shows that your preparation and/or attitude were not calibrated enough for what happened.

5 *Alert*. Similarly to above, when the situation seems unfavourable the brain can go into distress with a natural impulse for mourning. The feeling that something is going wrong is a great opportunity for leaders and managers to reach out and alert their colleagues (and themselves, of course) that a different course of action is needed. Keeping quiet during distress or trying to ignore the problem are the worse alternatives because very little can improve like that – if anything at all. Alert, as celebration, should always be a social response.

Recompose. Fear, or terror in the high end of this emotional state, is immensely beneficial when we face life or death situations since it focuses all our attention on the threat in front of us. Although fight-or-flight is the classic response to fear, accepting defeat and retreating fast are the most common ones suggested in Tomkins' model. Fear is a great tool for heightened attention but only for very brief periods of time. This is because it swiftly depletes many other vital brain functions since it redirects all energy to attention and muscle tension to respond to a potential deadly threat. Many managers who like fear, as in the opening case of this chapter, probably do so because it allows them to laser-focus themselves and the people around them on a task at hand. This comes with great side effects though and is a backfiring weapon when not used extremely carefully. Fear releases the steroid hormone cortisol, which, if triggered consistently, has devastating effects: killing brain cells, shutting down the immunity system and disturbing sleep cycles, among others (Brown *et al*, 2015). According to the authors of the book *The Fear-free Organization: Vital insights from neuroscience to transform your business culture*, 'it's like driving a car with one foot on the accelerator and one on the brake' (Brown *et al*, 2015). Recomposing ourselves and re-establishing our strategy by realistically examining threats with the help of those closest to us can be of great assistance when facing fear.

6 *Reassure.* Shame in Tomkins' model is more about hiding a feeling than expressing one. It is more about someone lowering extensively your joy rather than creating a totally new emotion. Leaders can face failure by retracting to their own selves, closing communication channels and deeming the situation helpless. Or they can use this emotion as a serious signal to review their behaviour and their overall stance in order to take appropriate measures for improvement. It takes willpower and a strong growth mindset to do but results can be spectacular. Reassuring yourself and your team of your collective abilities and plentiful opportunities for future success can work wonders for dealing with shame.

7 *Double-check.* Disgust is about having 'swallowed too much' that does not fit with our own values, attitudes and perspectives. It is an impulse reaction to reject things, people, ideas, situations we cannot digest, metaphorically speaking. Instead of showing contempt and alienating people, when feeling like instantly expelling someone or something it is far better to double-check where this response comes from. Is it justified or is it because of burnout-induced cynicism, stereotypes and other

biases, and misunderstood information? The same goes for dissmell, which is literally about pulling ourselves away from something as rapidly as we can because of its potentially toxic effect on our well-being. The impulse is to create as much distance as possible, as fast as possible. This emotion can take over when we momentarily consider a situation as 'toxic' and we want to automatically move back. Again, double-checking the reasons for this impulse is important to determine with a degree of certainty whether this is the best action (which it could easily be) or if we are just fooled by circumstances and our own biases.

All core emotions need to be understood and embraced by modern leaders. No emotion should be ignored, suppressed or voided since emotions have evolved over millennia to ensure our survival, both physically and socially. As Tomkins suggested though, our role is to increase the impact of the positive emotions over the negative ones. Utilizing all emotions, even the negative ones, for elevating our personal performance and the performance of our team is what separates inspirational leaders from managers who demotivate.

Action box: identify the emotion

Following the suggested model above, try to think of specific situations with intensive emotions that you have experienced recently. Try to write down some basic keywords that your brain could use in order to manage the emotions that emerge based on the key insights above. From now on you can use these key words in order to activate and deal with the emotions in different situations.

Elementary Dr Plutchik!

Understanding and dealing effectively with core emotions are the first necessary steps for becoming a better leader. The next, and more advanced step, is to understand how emotions are combined in order to create the wealth of feelings we experience in everyday life. In this direction, Dr Robert Plutchik, in the beginning of the 1980s, went beyond the identification of basic emotions, suggesting that each emotion, when combined with others, produces new ones and that each one forms the opposite of another emotion. By studying Dr Plutchik's emotional combinations we get a more complex but also a more representative view of the function of emotions in ourselves and others.

Starting with his view of the basic emotions, Dr Plutchik (2001) suggested that each of his eight core emotions has an opposite one: joy has sadness; trust has disgust; fear has anger; and surprise has anticipation. Knowing the opposite of an emotion is important, because the moment you experience a basic emotion you can tell which one is on the other side and take necessary actions. For example, when you do not trust a colleague enough and you want to improve the situation, you need to lower disgust that as we already saw leads to rejection and expulsion. Double-checking the facts, re-examining your perception and identifying in yourself the reasons by which this person's offering seems so opposite to your own well-being are a good start. Sharing aspects of your findings in a constructive and positive way with your colleague and working together to improve trust between you is the recommended way forward. The same goes for surprise and anticipation. Fast-moving changes in all industries and the huge impact of complexity in our working lives has made anticipation a difficult emotion to sustain. Constant anticipation can lead to certainty and certainty means security and control. Since these are in short supply in modern business environments, if we do not want ourselves and our people to be constantly surprised (an emotion that can easily develop into a negative one) then we should do our best to minimize creating rigid anticipation for specific outcomes as a precondition of well-being, personally and collectively (in the team). The small/purposive bets approach discussed in Chapter 3 can be very useful in avoiding big certainties and rigid anticipations. A strong sense of purpose, again from Chapter 3, can also help develop healthy and open anticipation and thus mitigate constant unpleasant surprises.

But Dr Plutchik went even further, suggesting that putting two emotions together creates new ones, and that knowing these ingredients of an emotion can help us deal with this emotion better. The main combinations are:

- Joy and trust create love (A). The opposite is remorse (B), created by sadness and disgust.
- Trust and fear create submission (C). The opposite is contempt (D), created by disgust and anger.
- Fear and surprise create awe (E). The opposite is aggressiveness (F), created by anger and anticipation.
- Surprise and sadness create disapproval (G). The opposite is optimism (H), created by anticipation and joy.

A scary boss who brings results usually leads through submission and fear, very much like in the opening case of this chapter. In our experience, submission unfortunately is dominant as a leadership emotion of choice in many organizations globally. Regardless of the advantages of obedience and predictability from strictly-followed rules, submissive behaviours can be extremely dangerous. It is not only the fact that creativity gets lethally wounded and that groupthink thrives in a submissive environment, but behaviours can easily take a darker turn too. As shown in the seminal experiments of Stanley Milgram and Philip Zimbardo, obedience can lead to immoral actions since people suspend their own moral and rational judgement under a submissive authority. Milgram (1974) conducted his experiment in the early 1960s, where participants, acting as teachers, were asked to cause pain to other participants, acting as students, by punishing them with electric shocks when they made learning mistakes. The students were actors who never received electric shocks and the electric shock-generator was faked, but the teacher-participants were ignorant of this. Still though, many teachers, up to 65 per cent, chose to administer electric shocks even of very high intensity, when the experimenter demanded from they do so in order to correctly follow the protocol of the test. In Zimbardo's (2007) infamous Stanford Prison Experiment conducted in 1971, participants were asked to live isolated in a prison environment and to act as guards or prisoners depending on the team they were assigned to. After just six days the experiment had to stop due to the extensive abuse that prisoners received both by guards and by fellow-prisoners on the guards' request. Those two classic experiments highlight that submission and obedience can easily transform some normal people into instruments of torture. It also demonstrates that the use of power/authority is something that by default involves emotions that make people change their attitudes towards others. In organizations, this means that combining fear and trust to lead people can get out-of-control and create an immoral environment in which cheating, lying and abusing become the norm. In a recent study various interviews with managers and employees from different companies operating in business environments that are under financial crisis were conducted (Psychogios *et al*, 2016). The results of this study show that managers are more willing to use the above behaviours in order to manage employee reactions to changes applied due to critical situation. In other words, the context itself (such as a crisis) can enhance submission and obedience as negative attitudes of leaders towards followers. As we have argued from the very beginning of this book organizations need engaged, passionate, creative personnel, which is what submissive behaviour can never help us develop.

Love, interestingly enough, is the combination of trust and joy, since excitement accompanied with faith in someone creates a strong bond. Love and submission both have trust as a key ingredient, but joy replaces fear to create love. This is a key point in how Plutchik's (2001) analysis of emotions can be of great assistance for leaders. Trust is crucial for developing followership, but accompanying it with fear or with joy can result in very different kinds of followers! We all need to dissect emotions and see the best combinations that can be used to motivate both ourselves and others.

> **Action box: decode Plutchik's emotional combinations**
>
> Defining and understanding emotions is not easy because, as Plutchik admitted, there are more than 90 definitions in relevant literature on what they actually mean. However, you can drastically improve managing your emotions by spending some time considering and applying his combinations. You can do this by taking each combination, from A to H, separately and trying to relate them to real-life examples from your own working environment. Try to see what will happen if you increase or decrease any of the basic ingredients. Most importantly, identify ways by which this is possible for you personally. For example, aggressiveness consists of anger and anticipation. If you feel constant aggressiveness towards a colleague, a department or a procedure is it possible to lower its negative impact on you by lowering or changing your anger, your anticipation or both? How would any of this emotions strategy be possible?

In taking the concept of emotional combinations one step further, the entrepreneur, speaker and author Chip Conley (2012) suggests a number of emotional equations, as he calls them, that can help us achieve our professional and life goals more efficiently and effectively. Following Plutchik's rationale of putting two emotions together to create something new, Conley created his own elaborate formulas. Three of them are congruent with our own viewpoint of modern, brain-based leadership.

1 Curiosity = Wonder + Awe. We have already discussed curiosity as a prerequisite for creativity as well as its importance in always asking questions, allowing the brain to process new information and challenge old biases. According to Conley, curiosity consists of two main ingredients, wonder and awe. While wonder is the pure

excitement of discovery, awe is fear and surprise combined. In order for our leadership curiosity to work well, we need to allow the pleasure of being faced with something potentially amazing, a new product/service, colleague, technology etc, to go hand-in-hand with humility and the feeling of being part of something bigger than ourselves. Awe allows us to connect with the world around us and adopt a more realistic view of the situation without losing the aspect of surprise when experiencing something grandiosely new. Curiosity without wonder is dry, shallow and short-lived, while curiosity without awe is full of arrogance, isolation and misinterpretation.

2 Regret = Disappointment + Responsibility. Within dynamic business environments leaders are expected to make many decisions daily. Although some of these decisions have higher weight than others, choices are usually plentiful for each decision and choosing one alternative over others can emotionally backfire. This equation suggests that the higher the disappointment, your responsibility, or both, the higher the regret. Regret itself, according to Conley, is not necessarily a damaging emotion but if it turns to remorse (extreme regret) then it can be devastating. We have seen through our experience that regretting a decision can be beneficial as far as it leads to deep learning and corrective future behaviours. However, we have also seen regret becoming the norm, driving managers to question their skills and damaging their decision-making ability. Regret in leadership is unavoidable. Minimizing your disappointments by adopting a long-term life view, searching for lessons in failure, delegating responsibility whenever appropriate and being always aware of the big picture are effective strategies to address this equation.

3 Thriving = Frequency of positive/frequency of negative. Thriving, or positivity as Conley also calls it, is about the relation between positive and negative events in our lives. Better put, it is about our perception of them. In order for this equation to have a beneficial outcome, positive events have to outnumber negative events by three to one. This is because negativity has a stronger pull on us than positivity, so positive events have to always be more than the negative for the equation to really work and to lead to thriving. Evolutionarily speaking, we have to develop a higher sensitivity to negative stimuli than to positive and this can be easily seen in the fact that in all models of core emotions the negative ones always outnumber the positive. In our primitive times, negative emotions ensured survival more than

positive ones and this is why we tend to notice the negative more than the positive. Yet, in today's safer environments it is positivity that leads to healthier and more sustainable motivation and thus to desired corporate behaviours that can ensure creativity, engagement and growth. Changing our perspective to start noticing more positive information around us, making sure we operate under the right values, appreciate any lesson from any event, and connect ourselves to other positive people can help us increase the nominator and decrease the denominator in this equation.

All these equations, together with Plutchik's emotional combinations, can have a positive or negative final effect on your leadership capabilities. It is your decision which one it will be. You have enough material now to start working on your emotional toolbox to create your own mix of emotional ingredients that will help you achieve personal and organizational objectives. Nevertheless, there is a specific positive emotional state that has been ignored by organizations for far too long. Although we all try to maximize it in our private lives, it seemed, and still seems for many, inconsistent with management. This is happiness and its advantages will surprise all those still believing that workplaces have to be charged with negative emotions such as fear for people to optimize their performance.

Bliss leadership

Claudia Hammond (2005), in her book *Emotional Rollercoaster*, mentions that happiness is usually at the top of the list when people are asked to name emotions and that it exists in almost all known models of core emotions (often though referred to as joy by scientists in order to capture more clearly its momentary expression). Based on this observation, it is remarkable that psychology and neuroscience were preoccupied for so long with the problems of our mind and not with its positive side. This is probably further proof of our brain's negativity bias as it was mentioned earlier. It was only in 1998 that the then incoming president of the prominent American Psychological Association (APA) Martin Seligman announced that psychology had to focus more on the positive side of our mind and called for his term's focus to be positive psychology (Wallis, 2005). It was about time that scientists' obsession with negativity gave way to a more balanced approach that included all those elements that make our mind thrive, not only those that make us mentally ill.

I can't get no... happiness

'I can't get no satisfaction,' the Rolling Stones have sung all around the world in one of their most recognizable songs, first released in 1965. And they were right at that time. Management, and marketing alike, considered satisfaction to be a core measurement for employees and customers respectively. The idea was that if employees, and customers, are satisfied they will behave in the way they should: employees with productivity and customers with purchases. This managerial idea found its limit in one of the most popular management paradigms that emerged in the '80s, called total quality management (TQM) (Psychogios, 2005). Satisfaction is mainly based on multiple discrepancies theories (Michalos, 1985), in which people's satisfaction depends on comparing themselves to multiple standards they hold including other people, goals, ideal levels of satisfaction and past conditions. Satisfaction is higher when the comparison shows favourable results (downward comparison – set standards are lower than current personal state) and lower when comparison is unfavourable (upward comparison – set standards are higher than reality). The problem with this approach is that satisfaction alone cannot lead to the heightened behaviours we want ourselves and our employees to demonstrate. It seems that satisfaction has become, most of the time, a necessary but not a sufficient condition for increased engagement and boosted loyalty. In numerous studies we have done in our clients' organizations, using both quantitative and more exploratory qualitative methods, to help them identify key issues for performance improvement, it constantly came up that satisfaction was rarely related to any important factors. We believe that satisfaction has a new meaning today, mainly because of uncertainty. The latest global financial crisis, increased competition from around the world, fast technological advancements, corporate outsourcing and downsizing programmes, and even climate change creates a landscape of the unknown that naturally makes many people feel unsafe and insecure. In such an environment, just having a stable job and getting paid regularly can be, for many people around the world, a good reason for being satisfied. It actually means 'I am OK at the moment, look around you' or 'I am OK at the moment, I do not have bigger ambitions' rather than 'We will do great!' Furthermore, and following Herzberg's renowned hygiene and motivation theory (Herzberg *et al*, 1959), we are observing that satisfaction itself has become a hygiene factor rather than a motivational one. This means that satisfaction as a measurable factor and explicit

> business terminology on employees' well-being is actually measuring basic acceptance at best and nothing more.
>
> No great leader will be ever remembered because of creating a satisfying working environment. Satisfaction is not enough. Great leaders are great because they can spread strong and heart-quickening, positive emotions. So, it is Pharell William's global hit song 'Happy', released in 2013, that forms the soundtrack of modern leadership and not the Rolling Stones' 'Satisfaction'.

Moving towards a more positive view of the mind was necessary to start taking seriously, and to unlock the secrets of, positive emotions. Happiness (or joy, or subjective well-being among other names) quickly rose to become the positive emotion of choice in both science and pop culture. This unfortunately means Smilies everywhere. But fortunately it also means significant research into the inner workings of happiness in our brains and behaviour. For both the wider society and organizations, the study of happiness brought a surprising and powerful insight. Happiness is not just an outcome of successful work, but a prerequisite. This counterintuitive finding went against existing values since in many cultures around the world one is entitled to happiness only after success is secured beyond any doubt. However, Boehm and Lyubomirsky (2008) have found substantial evidence in support of the reverse hypothesis, that is, happiness is the reason why some employees are more successful than others. By conducting a review of a large number of cross-sectional, longitudinal and experimental studies they concurred that:

> Taken together, the evidence suggests that happiness is not only correlated with workplace success but that happiness often precedes measures of success and that induction of positive affect leads to improved workplace outcomes.

This means that inducing a positive atmosphere at work will not undermine success, but it will actually cause it. Withholding positive feelings just for when success comes along will inevitably lower the chances that success will ever come. And the evidence for this effect is widespread. Simon Achor, the global advocate for happiness at work, summarizes relevant research in his best-selling *The Happiness Advantage* book (2010), showing that:

- Optimistic salespeople do 56 per cent better at their work than pessimistic colleagues.
- Doctors in a positive mood show three times more intelligence and they are 19 per cent faster in diagnoses than others.

- Students primed (put in a mental state without knowing) to feel happy before taking exams did far better than those primed to feel neutral.

Such strong evidence, along with his own empirical research on more than 1,600 high-achieving undergraduates at Harvard and his professional engagement with top companies around the world, lead him to claim that:

> It turns out that our brains are literally hardwired to perform at their best not when they are negative or even neutral, but when they are positive. Yet in today's world, we ironically sacrifice for success only to lower our brain's success rate. Our hard-driving lives leave us feeling stressed, and we feel swamped by the mounting pressure to succeed at any cost.

And this cost, it seems, is success itself. There is a very interesting surprise here. In difficult times, when we personally are and/or the company we work for is, not doing very well, we allow the negative emotions to get hold of us and we postpone happiness, individually or collectively, for when things get better. This is exactly the happiness-averse mindset that makes success more difficult and more remote. What we need to do is the opposite. Regardless of the current conditions, we need to retain our high spirits and always strive for the best. This, mathematically, can lead to success more easily than negative or neutral emotions that we might find more fitting for difficult times. This is why we always try to create the best possible uplifting conditions when we work with a client and also to keep ourselves constantly upbeat. This is also supported by a recent study about organizations that are hit by financial crisis. In particular, by investigating various case studies of companies that struggle to survive in a high crisis environment, through interviewing managers and observing behaviours and managers' and employees' interactions, it was found that in the cases where managers and leaders focused on the positive outcomes and emotions they achieved better results and ensured the survival of their companies during the turbulent period of time (Psychogios and Szamosi, 2015).

Happiness sets the foundation of success in any project because of its ability to release intelligence, creativity, collaboration and commitment. We distinctly remember a situation when a team that was deemed ineffective sprang into frenzied action when a positive atmosphere was created after our recommendation and interaction with the team members. At the end of the year this initially forsaken team received the highest achievement award by the company's CEO. No person, no team, no company can categorically be called ineffective before a positive approach is wholeheartedly and thoroughly applied.

But how can we work towards creating more happiness at work for us and others in order to calibrate our brains for success? Professor and author Richard Wiseman, after reviewing the available literature on improving happiness, concluded that the fastest and surest route to a happiness boost includes the following three groups of actions (2009):

1. *Gratitude and appreciation.* Happiness starts from the inside out and thus it needs a sunny place to start from. Create this sunny place by periodically expressing gratitude for everything good that has happened to you. If possible, create a written account of them. Frequently refresh your memory of fantastic experiences you've lived and how they made you feel at that time. Try to recapture and relive the sensation within your mind and body. Examine your current situation and handpick only what is positive. Try a bit harder and you will be surprised by how many positive moments go unnoticed because of the overwhelming and sticky power of negativity. Put aside the mind's negative veil and sunrays will shine through. Finally, do not hold back your appreciation for those close to you. Actively thank them for their contribution and efforts as well as for their personal support to you. When showing appreciation try to be as specific and concrete as possible. This will benefit both you and the person you are showing appreciation to.

2. *Experience and sharing.* Our brains are hardwired to react to real-life experiences better than to our material belongings. So, in order to boost your happiness fast, start engaging in refreshing and rewarding experiences as soon as possible. Also, create and offer such experiences to your team, not just material rewards. The 2015 results of *Fortune*'s Top 100 Best Companies to Work For (Colvin, 2015) suggested that free perks and futuristic offices are not the decisive factors for a company appearing on the list but rather the ability 'to foster strong, rewarding relationships... among their employees'. So, it is the intangible nature of the amazing experience of working with the people you like that makes people happy in their workplace and not the tangible benefits. It is only natural then that sharing has been scientifically proven to be another strong booster of happiness. Regardless of most people believing the opposite, sharing an item or an experience makes us happier than when we keep things only for ourselves. As Professor Wiseman reports, even five non-financial acts of kindness per day

can substantially increase our happiness. No surprise then that Random-Acts-of-Kindness (RAKs) has become so popular in management publications worldwide.

3 *Body language and behaviour.* Whenever we perform in class the famous happiness-inducing experiment, which includes holding a pen with your teeth, the results are astounding. If you hold a pen horizontally in your teeth without touching it with your lips for few seconds your brain immediately releases happy hormones just because it thinks you are smiling. Since you are smiling, something good must be happening! On the contrary, if you hold the end of the pen with your lips forming an O and not touching it with your teeth, your brain believes you are frowning and releases stress hormones. The amazing finding from this classic experiment is that happiness is sustained after the end of the test and makes people connecting more positively and improving their memory of happy events. This remarkably shows that there is a two-way communication process between the brain and the body. It is not only 'what the brain feels the body will show' but the other way around too. Smiling more, adopting a more positive, confident and upright body posture, using more positive wording when speaking and generally acting more happily will tell the brain to behave accordingly. Paraphrasing the adage 'be the change you want to see' we advise you to 'be the happiness you want to feel'.

Connecting the emotional dots

From an emotion-based point of view, if fear and negativity were Motivation Version 1.0, and happiness and positivity Motivation Version 2.0, there is a Version 3.0 that calls for a holistic approach to utilizing emotions in our personal and professional lives. Regardless of the model, all core emotions have strong evolutionary reasons to exist. The mere fact we have them today proves that they were instrumental to getting us to where we are now as humans. We need though to be able to recognize them whenever possible in order to determine the benefits or dangers that derive from them. As Plutchik (2001) eloquently put it, 'emotions can sometimes fail in their adaptive tasks'. This means that not everything we feel is justified by real circumstances and not all that we are inclined to do, pushed by our emotions, is beneficial for us and the people around us.

As the science of optimism, happiness and positive psychology came to improve the old simplistic models of behaviour such as 'the carrot and the stick' and unlock the huge benefits of having a positive outlook in life and work, more advanced approaches started emerging which will eventually improve the management of emotions. Todd Kashdan, a renowned expert on negative emotions, and Robert Biswas-Diener, a famous positive psychologist (2015), are advocating wholeness as the key to emotional success. As they argue:

> ... [I]t is high time that we reevaluated long-held beliefs of what is negative and what is positive, psychologically speaking. It is time for a new way of understanding of what it means to be mentally healthy and successful, to see both positive and negative as part of the larger, and more viable whole. This, then, is the Holy Grail of psychology, wholeness... Wholeness is to psychology what enlightenment is to spirituality.

They see signs of wholeness already appearing in corporate language when leaders talk about full engagement and optimal performance. Such language reveals that organizations prefer emotional maturity and utilization of any emotion from simplistic, rigid and out-dated categorizations of emotions to good/positive and bad/negative ones. For them, it is only wholeness that can lead to real emotional agility, which is not about avoiding negative emotions but about 'taking the negative out of them'. They cite Adler and Hershfield's (2012) study of 47 adults, which showed that those people experiencing psychotherapy happiness at the same period as experiencing sadness, had better results on their well-being than those who were experiencing the one or the other emotion only. The study concluded that changes in mixed, or concurrent, emotional experiences are a prerequisite for overall emotional improvement. Simply put, one has to experience, and deal with, both positive *and* negative emotions rather than prioritizing one over the other as in the cases of Motivational Versions 1.0 and 2.0, which preferred negative and positive emotions respectively. Such results are ground-breaking since they shatter the barriers between emotional categories long held in psychology as explained earlier in this chapter. Yes, we do have emotions that can be generally called positive and negative, and neutral for that matter, but this should not lead to stereotyping them as good and bad. As we mentioned in the section 'Applying Tomkins' model in organizations' earlier in this chapter, each of the basic emotions has to be experienced and even expressed in appropriate ways in order to utilize them more effectively and optimize their benefits. Each emotion gives us important signals about the environment and our own acclimatization to it, which we should never ignore and/or suppress. Leaders applying wholeness in emotions know what to keep and what not to keep from every emotion and above all, understand the value of all emotions to their motivation, flexibility and well-being.

A deeper and more scientific view of emotions is necessary for leaders to perform better but it is also necessary for our society to improve as a whole. In this direction, even animated movies can help. *Inside Out*, created by Pixar Studios and released by Walt Disney Pictures in 2015, introduced the science of basic emotions to kids and parents alike. In the movie, the basic emotions of joy, anger, sadness, disgust and fear are personified in the heads of the main characters. This personification works well in showing how emotions are the key sources of individual behaviour as well as in portraying the crucial and mutually dependent role of emotions in interpersonal relationships. In a world where the heart *emoji*, a highly emotional ideograph, was proclaimed as the top English word of the year for 2014 in online usage by the Global Language Monitor (Gander, 2014) such wide popularization of science can have a very positive effect on our collective understanding and acceptance of the complexity and beauty of our emotional brain.

Boost your brain: identifying motivational versions in the workplace

Create a table and put the three motivational versions as columns. In separate rows put all superiors, colleagues and employees that are positioned 360 degrees around you. Examine name by name and characterize each of them as high, low and medium for each of the motivational versions. The version that is dominating a person should be scored as high, the one that appears from time to time as medium and the one that is almost or completely missing as low. Remember that:

Version 1.0: Fear and negativity.

Version 2.0: Happiness and positivity.

Version 3.0: Wholeness and emotional agility.

While considering every person, try to remember specific instances that, when accumulated, indicate specific patterns of behaviour as a result of the preferred motivational version. At the end of the list add your own name and do the same for you. Try to be as open as possible. Then ask at least two people you trust in the organization to rate you by explaining the concept to them. Request that they provide specific instances to support their judgement. What did you learn from the whole exercise? Do you or the people around need to change their version? In what way? And how?

Keep in mind

Carrots, sticks, satisfaction and other single-emotion approaches are a thing of the past, incompatible with modern neuroscience. Brain chemistry and circuitry, as well as the close study of physical expressions, suggest that we have a number of core emotions that have evolved to perform significant tasks in order to increase our survival chances. Leaders and managers cannot perform their duties without understanding how these emotions work and for what reasons. Familiarizing yourself with the number and nature of our basic emotions, unlocking the surprising relations between them, taking advantage of emotional equations, utilizing the power of positive psychology and finally putting together the entire puzzle of emotions with wholeness and emotional agility are essential for brain adaptive leadership. Above all, our brain is an emotional organ.

References

Achor, S (2010) *The Happiness Advantage: The seven principles of positive psychology that fuel success and performance at work*, Crown Publishing, New York

Adler, JM and Hershfield, HE (2012) Mixed emotional experience is associated with and precedes improvements in psychological well-being, *PloS one*, **7** (4), URL: http://journals.plos.org/plosone/article?id=10.1371/journal.pone.0035633, accessed 5 July 2015

Badawy, AAB (2003) Alcohol and violence and the possible role of serotonin, *Criminal Behavior and Mental Health*, **13** (1), pp 31–44

Boehm, JK and Lyubomirsky, S (2008) Does happiness promote career success?, *Journal of Career Assessment*, **16** (1), pp 101–116

Brown, P, Kingsley, J and Paterson, S (2015) *The Fear-free Organization: Vital insights from neuroscience to transform your business culture*, Kogan Page Limited, London

Colvin, G (2015) Personal bests, *Fortune*, 15 March, pp 32–36

Conley, C (2012) *Emotional Equations: Simple truths for creating happiness+ success*, Simon and Schuster, New York

Ekman, P (2007) *Emotions Revealed: Recognizing faces and feelings to improve communication and emotional life*, Owl Books, New York

Fox, E (2013) *Rainy Brain, Sunny Brain: How to retrain your brain to overcome pessimism and achieve a more positive outlook*, Arrow Books, London

Gander, K (2014) Top words of 2014: The heart emoji named most used term of the year, *The Independent Online*, Monday, 29 December, URL: www.independent.co.uk/news/weird-news/top-words-of-2014-the-heart-emoji-named-most-used-term-of-the-year-9948644.html, accessed 19 June 2015

Hammond, C (2005) *Emotional Rollercoaster: A journey through the science of feelings*, HarperCollins, New York

Heiss, ED (2014) The MBA of the future needs a different toolbox, *Forbes online*, URL: www.forbes.com/sites/darden/2014/10/01/the-mba-of-the-future-how-many-doing-what/, accessed 6 August 2015

Herzberg, F, Mausner, B and Snyderman, B (1959) *The Motivation to Work*, 2nd edn, John Wiley, New York

Jack, RE, Garrod, OG and Schyns, PG (2014) Dynamic facial expressions of emotion transmit an evolving hierarchy of signals over time, *Current Biology*, **24** (2), pp 187–192

Kashdan, T and Biswas-Diener, R (2015) *The Power of Negative Emotion: How anger, guilt and self doubt are essential to success and fulfilment*, Oneworld Publications, London

Lövheim, H (2012) A new three-dimensional model for emotions and monoamine neurotransmitters, *Medical Hypotheses*, **78** (2), pp 341–348

Michalos, AC (1985) Multiple Discrepancies Theory (MDT), *Social Indicators Research*, **16** (4), pp 347–413

Milgram, S (1974) *Obedience to Authority: An experimental view*, Harpercollins, New York

Nathanson, DL (1992) *Shame and Pride: Affect, sex, and the birth of the self*, WW Norton, New York

Panksepp, J and Biven, L (2012) *The Archaeology of Mind: Neuroevolutionary origins of human emotion*, W W Norton & Company, New York

Plutchik, R (2001) The nature of emotions, *American Scientist*, **89** (4), pp 344–350

Psychogios, AG (2005) Towards a contingency approach to promising business management paradigms: The case of total quality management, *Journal of Business and Society*, **18** (1/2), pp 120–134

Psychogios, A and Szamosi, TL (2015) *Fight* or *Fly*? Rationalizing working conditions in a crisis context, 31st European Group of Organization Studies (EGOS) Colloquium, Athens, 1–4 July (Conference Proceedings)

Psychogios, A, Prouska, R, Szamosi, TL and Brewster, C (Forthcoming 2016) Exploring managers' and employees' experiences of working life within different crisis contexts: Lessons from Greece and Serbia, *International Journal of Human Resource Management*

Schultz, W (2002) Getting formal with dopamine and reward, *Neuron*, **36** (2), pp 241–263.

Tomkins Institute (2014) Nine affects, present at birth, combine with life experience to form emotion and personality, *The Tomkins Institute Online*, URL: www.tomkins.org/what-tomkins-said/introduction/nine-affects-present-at-birth-combine-to-form-emotion-mood-and-personality/, accessed 14 July 2015

Tomkins, S (2008) *Affect Imagery Consciousness* (Volumes I and II), Springer Publishing Company, New York

Wallis, C (2005) The new science of happiness, *TIME Magazine*, **17** (1), URL: http://content.time.com/time/magazine/article/0,9171,1015832-1,00.html, accessed 10 July 2015

Wiseman, R (2009) *59 Seconds: Think a little, change a lot*, Macmillan, New York

Yu, AJ and Dayan, P (2005) Uncertainty, neuromodulation, and attention, *Neuron*, **46** (4), pp 681–92

Zimbardo, P (2007) *The Lucifer Effect: Understanding how good people turn evil*, Random House, New York

SUMMARY OF PILLAR 2: FEELINGS

TABLE S.2 Summary of pillar 2

Continuously improve emotional style	Improve your self-perception of emotions by: • describing in detail the situation; • giving a detailed account of the behavioural response that the feeling produced; • creating a list of events that had similar behavioural outcomes; • looking inside you with minimum interruption and with genuine interest; • listening to others carefully and trying to understand their viewpoint; and • scanning constantly internally and externally for emotional cues.
Be aware of moods	Keep yourself on the positive side and increasingly on the high energy. Consider your mood often to make sure you give yourself the chance to experience the right emotion at the right time.

TABLE S.2 *Continued*

Be aware of the power of the basic emotions	There are nine basic emotions that leaders need to be aware of: enjoyment, interest, surprise, anger, distress, fear, shame, disgust and 'dissmell'. Deal with the core emotions through: • celebrating any success for yourself and your team; • exploring new data, people, situations and other possibilities; • stopping to think, especially when experiencing unexpected changes; • recharging when you are full of anger by considering if this is the best way forward; • alerting yourself and your colleagues when a different course of action is needed; • recomposing yourselves and re-establishing your strategy by examining threats; • reassuring yourself and your team of your collective abilities and opportunities; and • double-checking the reasons for creating as much distance as possible.
Develop emotional agility	Develop emotional agility by following these three groups of actions: • gratitude and appreciation; • experience and sharing; • body language and behaviour.

PILLAR 3
Brain automations

Gut reaction, faster solution

06

I am in control, I will change

He is a revered and fearless negotiator. He always has been. His advanced skills for closing the right deal during the most difficult circumstances got him his current position as the CEO, his first time at the top position, in a mid-sized investment fund. The fund wanted someone who could quickly find the right pathway to growth, since the last financial crisis had driven it into problems that would not go away easily. Determined to succeed, he started applying everything he had learned in his career from day one in dealing with both external and internal people. He renegotiates the already existing deals, identifies new opportunities and quickly establishes valuable connections. He makes progress fast on the inside too, turning around bad decisions of the past, cutting costs and streamlining operations. His early achievements are widely celebrated by shareholders, who trust him almost completely. However, his quick wins soon turn sour.

His ability to negotiate by quickly finding the opponent's weaknesses and attacking with surgical precision served him well in task-specific, highly-specialized jobs of the past and in setting the pace early in his new one. However, this same ability started creating more problems than it solved. With the passing of time it became evident that his shark-like approach to problem solving backfires in developing and maintaining relationships. Many of his top people within the company resigned and changed jobs, and many of the remaining ones are thinking of doing so shortly. Although he seems to perform very effectively and efficiently in the technical parts of the job, he somehow misses incorporating the human factor as effectively. When confronted by some of the shareholders on this he claims the opposite: it is because of his deep understanding of human beings that he can be so good in negotiations and in getting what he wants. He knows exactly how to discover soft spots, such as fears and desires, and how to press them in order to achieve the outcome that he wants. He acknowledges the significant role of emotions in the whole process, both in him and in others, and this is why he can get the numbers he strives for. He is not cold-blooded or cool-headed: he just thrives on manipulation.

One morning, his closest ally in the company, and his own recruitment choice a year ago for deputy CEO, resigned. The CEO, shocked from the news, calls his deputy for urgent discussions. During a two-hour heated conversation the ex-deputy told him that he can no longer work alongside him, because he simply cannot tolerate the premeditated manipulation he exerts to himself and others. 'People should be treated with respect and not with a different approach in each moment in order to get what you want from them,' he yelled. Changing faces, emotions and words so easily, day in and day out, to achieve results is not something that is valued in leaders, he continued. The CEO, agreeing on these observations, pledged to change. He would do his best to treat people differently, be more consistent in his attitude towards others and put values over short-term gains. It was about time someone told him so and he is determined to be a better person, a better CEO, a better leader. He considers himself a very mentally strong person, in full control of his behaviour, and this is what he is going to do. He convinced his deputy to stay.

Two months later, however, the deputy is gone for good. The CEO never really changed, regardless of his commitment and even initial sincere efforts. The irresistible pull of his deeply rooted behaviour proved much stronger than his understanding of the need to change. A year after his deputy left, his initial great results turned to losses and the shareholders very seriously consider the option of finding a new CEO.

This case shows that the concept of consciousness in human thinking and its role in what actually makes us human is central to neuroscience and psychology. Do we take conscious decisions fully aware of risks and benefits? Does our mind affect our behaviour directly and exclusively? Can we change whenever we want to? Do we make our own fate? Are we in control? If yes, then we only need to train our conscious minds through formal and informal learning processes, as we have been doing for centuries. But if not, then leaders need to look deeper to find effective ways to understand and influence better their decisions and actions. Because as we now know, we are not in control of ourselves, at least to the extent that we believe we are. Our brain is.

The mind-controlling brain

'The unconscious is the part of the mind that contains feelings and thoughts that we are not aware of and it influences the way we behave' (*Cambridge Advanced Learner's Dictionary and Thesaurus*). The power of the unconscious has long been emphasized in psychology, with Sigmund Freud pioneering our understanding of how our deepest and most unknown desires affect our everyday lives. However, most recent research in the field reveals that the unconscious is not the dark place that Freud envisioned. On the contrary, it brings about unique advantages. Its processing capacities form an integral part of our healthy, dynamic and interactive state of being. McGowan (2014) highlighted that 'the nature of the unconscious thought that emerges from contemporary experiments is radically different from what Freud posited' and that viewing the mind with its unconscious influences 'honors the unique experience of individual human beings – something often overlooked by the current medical approach to the mind'. Our unconscious seems to be our biggest strength as species and not our darkest weakness.

The impact of the unconscious on our decision making is profound – much more than we are willing to admit. But it is there and, by ignoring it, we can only harm our leadership potential. In the most famous experiment on the primacy of the unconscious on decision making, Benjamin Libet and his team asked participants to make simple decisions, such as pressing a button or moving their fingers whenever they felt like it, within a timeframe (Libet *et al*, 1983). They also asked them to note their decisions at the time they were made during the experiment. By monitoring participants' brain activity with an EEG (electroencephalograph) and their voluntary motor activity with an EMG (electromyograph), Libet and his associates revealed that the electric build-up on their brains for the upcoming decision appeared approximately a quarter of a second earlier than the moment when participants became conscious of that decision, and almost half a second before they acted. This *readiness potential* in our brains comes before we are aware of our own decisions. And if you think this is shocking, Soon *et al* (2008) have more recently raised this gap to 10 seconds for simple decisions, claiming that this delay 'reflects the operation of a network of high-level control areas that begin to prepare an upcoming decision long before it enters awareness'.

Our brain prepares to take a decision and nudge us into specific actions much earlier than when we become aware of it. And it does this for very good reasons. The unconscious helps us save energy (as we have seen in Chapters 1 and 2), react very fast in life-threatening situations, and form attitudes fast about others based on previous experiences. Our brain's unconscious processes underlie the whole way we deliberate and plan our lives (Bargh, 2014). In his best-selling book *Blink: The power of thinking without thinking*, Malcolm Gladwell (2005) compares our *adaptive unconscious*, as it is known, to a giant computer that quickly and quietly processes a lot of data we need in order to keep functioning normally in our everyday lives. If we processed all this stimuli consciously we would be barely able to take any action at all. Gladwell quotes the renowned psychology professor Timothy Wilson, who coined the term adaptive unconscious, saying:

> The mind operates most efficiently by relegating a good deal of high-level, sophisticated thinking to the unconscious, just as a modern jetliner is able to fly on automatic pilot with little or no input from the human, 'conscious' pilot. The adaptive unconscious does an excellent job of sizing up the world, warning people of danger, setting goals, and initiating action in a sophisticated and efficient manner.

In support of this notion, psychologist and behavioural expert Gerd Gigerenzer (2007) has also called for a wider acceptance of the *intelligence of the unconsciousness*, which is fundamentally based on simple rules of thumb, which in turn are based on evolved capacities of our brain. These evolved capacities help us make snap decisions or fast judgements and to act accordingly, by providing signals in the form of gut feelings. Taking into account that our advanced language ability is positioned in the top part of our brain, the cerebral cortex, deeper and more ancient brain systems try to communicate with us through gut feelings, impulses, instincts and intuitions, since they simply cannot speak. And this deeper, adaptive, intelligent unconscious has many times been proven to be able to make better decisions in situations as demanding as choosing stocks to invest in and buying a new property (for more information, see Wilson 2002, Gladwell, 2005 and Gigerenzer, 2007).

This is not to say that unconscious thinking is always superior to the conscious one and that we have to abandon analytics altogether. The scientific debate is ongoing with both sides conducting experiments to show that trusting your unconscious mind is either more beneficial or more dangerous in complex decision making, where there are a lot of factors determining the outcome and, even worse, most of them cannot be easily understood or cannot be understood at all. On the one side, Ap Dijksterhuis, Loran Nordgren and associates, supporting their Unconscious Thought Theory, are claiming that unconscious thinking, or deliberation-without-attention, can be far more beneficial than conscious thinking when a problem is very complicated and multifaceted. In such situations, too much thinking leads to bias activation, analysis-paralysis, worse decisions and considerable less post-decision satisfaction (Bos *et al*, 2006). On the other side, scholars such as Ben Newell (Newell *et al*, 2009) claim the opposite, saying that only extensive rational thinking can bring results within complexity while fast, unconscious or unfocused thinking can lead to serious mistakes. Regardless of where you personally stand in this debate, the important point is that decisions do not just form from thin air. The brain spends considerable energy before a decision is made in order to prepare the conscious mind for it. The quality of this preparation will directly influence the quality of the decision.

We usually do a very simple experiment when we want to demonstrate to our audiences the power of the unconscious will over the conscious one. We ask them to try *not* to think of a white polar bear for the next five minutes, which is almost impossible for most of us. Sooner or later a polar bear appears in our mind's eye. In fact, although we try to avoid thinking of a white polar bear by filling our conscious mind with other, different thoughts, our unconscious mind remains alert for any signs of the unwanted thought, the better to help us chase it away. This popular thought exercise, first mentioned by the author Fyodor Dostoyevsky in 1863 and proved in experiments by Daniel Wegner of Harvard University (Winerman, 2011), helps us to highlight the fact that our conscious mind is not as in control as we thought – if it is at all. Our unconscious is. The good news is that we can study the way the unconscious mind works and recalibrate it for the benefit of further developing our managerial and leadership skills.

Action box: experiment with unconscious and conscious thinking

Select an organizational problem of high complexity (a lot of factors known or unknown involved) that your company would like to solve in the near future. Take your team and separate them into three smaller groups. Ask Group A to take a look at the problem and make an immediate decision. Spending time in thinking of any kind is not allowed here. Ask Group B to do the opposite. They can take the data, add some more if needed, spend half an hour at least in advanced deliberations and then make up their minds. Ask Group C to consider the problem briefly, but then distract them with any trivial game of your choice (solving anagrams, puzzles, angry birds, etc). They have to decide about the complex problem of your organization after they finish playing the game without further deliberation. You can even add a Group D that has to literally sleep on the problem overnight and suggest a solution in the morning, without having any prior interaction or deliberation on the topic.

Observe any differences in the decisions made. Which decisions would you follow and why? Is it the time of rational deliberation that determines the quality of the outcome or something else?

Reshuffle the groups and repeat the experiment with a problem of very low complexity (a simpler issue with fewer aspects and attributes). Do you see any differences between the two cases?

In the first case (complex problem) rational thinking with over-analysing all parameters of the problem, it may end up with brain paralysis since the nature of the problem (complex) involved many unseen factors that cannot be easily detected by the conscious mind and therefore we need more intuition and activation of the unconscious mind. In the second case (simple problem) the conscious mind can easily take control since factors can be seen, explored and determined accordingly.

It's prime time

Automatic brain responses to real-world situations are the norm rather than the exception in our everyday lives. Taking into account that 98 per cent or more of our total brain activity daily is completely unconscious (Gazzaniga, 1998) and that 95 per cent of our decisions are unconscious (Zaltman, 2003), this should not be a surprise. What is surprising though is the fact that many managers and leaders go about their careers being either oblivious or conceptually against the idea that their behaviour, and that of their colleagues, is highly driven by the unconscious. They need to understand this is true and that this is not necessarily a bad thing.

We mentioned above the phenomenon of the brain's *readiness potential* in the process of making a decision. Before we consciously make up our minds and move to a behaviour, our brain's neurons start firing within the unconscious realm, creating the basis for this decision to be made. This neural preparation process for decision making is not independent from the external environment and can be highly influenced by stimuli. Influencing the readiness potential of our brain means influencing our decision making and behaviour, so that we can direct or nudge our actions in a preferred path. The way this is achieved is called *priming*. Priming is an effect of our implicit memory in which exposure to one stimulus affects the response to another one (Meyer and Schvaneveldt, 1971). Priming is preparing our brain for specific thoughts and behaviours (Alter, 2013). The brain will unconsciously and very quickly consider cues that are detected in a situation to prepare our mind to make a suitable decision and to adopt a fitting behaviour. Different external cues will lead to different neural priming/ preparation and possibly to different decisions and behaviour. In essence, messages of the environment can put our brains in specific thinking and acting pathways. In order to optimize our own performance and that of our team we need to make sure that we use priming in a positive and productive way. Otherwise, priming can work heavily against us, without us ever knowing it. The list of experimental results indicating the power of priming is long (McGilchrist, 2012):

- Participants in a general knowledge quiz were separated into three groups, each one engaging in different activities that would prime their brains into different mental states: one as professors, one as secretaries and one as hooligans. The professor-primed group significantly outperformed the other two in the quiz, with hooligans doing worse of all.

- Participants primed with a mental state of elderly people (reminded that the elderly are sentimental, playing bingo and have grey hair among other characteristics) became more conservative in their opinions while others that were primed with a politician state of mind (reminded of the main characteristics of politicians) expressed themselves in a lengthier manner.

- When elderly people were primed with the positive association of being old, they performed much better in memory tests than those primed with the negative aspects of being old.

And that's not all. In a much publicized study, US voters participating in a priming experiment on American identity and willingness to vote at San Diego State University in 2007, considered the then US presidential candidate Barack Obama as less American than even Tony Blair, the former British Prime Minister! They were initially primed to consider Mr Obama's ethnicity, labelled in the experiment as 'black' (Kristoff, 2008). Such negative racial biases can sway people's opinions in subconscious ways and thus need to be confronted openly in order to lose their undetected power over our decision making.

In a series of three famed experiments, Bargh and associates (1996) found that students:

- Primed by sentences with associations to the elderly population, walked slower in the corridor after they were primed than those of a control group that read non-associative sentences.
- Primed by sentences with associations to polite and respective behaviour, waited more patiently and for a longer time outside the office of a busy researcher (who was part of the experiment) than those primed with words such as 'rude', 'disturb' and 'aggressiveness'. Actually, 82 per cent of those primed for politeness did not interrupt the busy researcher at all.
- Primed by words that reminded them of their race, African American students reacted with more hostility to a researcher's frustrating requests. Again, as in the case of the Obama study, negative biases in issues of race (even 'self biases' observed in this particular study) can lose much of their subconscious influence when people become aware of them.

All these results, and they are all worth noting, demonstrate that our brain automatically responds to external cues by subliminally adopting a specific frame or state of mind. Priming prepares the brain to take the 'right' decision and go into the most 'appropriate' behaviour by evoking attitudes and models pre-established in our long-term memory.

Priming works best when people are not aware of its effect. When we become aware of the effect, priming stops working. This means mainly two things for leaders. First, when we detect in ourselves behaviours or reflex reactions that we do not like or prefer we should seriously consider whether they are the product of unfavourable priming to our brain. Are there any external cues that are automatically placing us in an unproductive and ineffective mental state? Second, we should do our outmost to participate in and/or create a working environment that is priming our brain with the best attributes for high performance and the highest chances of success. Are the companies we work for, our colleagues, even our offices boosting our brains for maximum performance or do they do the opposite? The core takeaway though is that our brain is not independent from its environment. It will continuously receive and process subliminally much more information than our conscious mind could ever handle and it will reflexively respond by altering our conscious mind's ability to make decisions. Neural readiness potential and mental preparation for decision making through priming are what our brains have evolved to do very well. Leaders ignoring or opposing those scientific insights are just missing the amazing opportunities of recalibrating their brains for greatness. This is because the knowledge of the unconscious mind's influence on our behaviour is the single most important weapon we have to shape it favourably for us. Knowledge is true power, in this respect. As Bargh (1994), a prominent advocate of our unconscious mind whose research and words are included prominently in this chapter, stated categorically on the matter:

> Automated social cognitive processes categorize, evaluate and impute the meanings of behavior and other social information, and this input is then ready for use by conscious and controlled judgment... [but] the unintentional and uncontrolled nature of automatic analyses of the environment does not mean they are impossible to control or adjust for when one is aware of them, if one desires.

Our unconscious mind is more powerful than our conscious one, in the sense that it consumes more brain power and analyses more information than our controlled thinking process. The challenge we all face in our jobs is not to find ways to stop this power, a very dangerous and neurologically almost impossible thing to do, but to harness it to our advantage.

Priming, like other influential concepts in science, has not been without opponents. Some experimental results failed to replicate in later studies (see Doyen *et al*, 2012 and Bargh, 2013), bringing priming under fire in an outspoken debate. In particular, these studies claimed that their experiments failed to show priming, suggesting that both priming and experimenters' expectations are instrumental in explaining social behaviour. The latter arguments triggered the renowned psychologist Daniel Kahneman (2012) to issue a public letter to fellow researchers in support of more research in priming, a concept he claimed he believes in. Even Doyen and associates, who were among the ones who did not manage to replicate priming results in the exact same way as in one of the famous studies, explicitly concluded that 'unconscious behavioural priming is real' (Doyen *at al*, 2012).

Automated cognitive processes, priming and readiness potential are concepts that are here to stay since our unconscious mind seems to be in higher – and more beneficial – control of our thoughts and behaviours than we ever expected. Having said that, we are not simple zombies in the service of our uncontrolled brain. Through awareness and knowledge we can decrease the negative effects of priming, such as stereotyping, and increase its positive ones in order to maximize our leadership and managerial performance. Nevertheless, those that still oppose the presence, power and often intelligence of our unconscious mind, hailing the unconditional superiority of our rationality, will be living in an illusion that is both ineffective and hazardous for all. Like in the opening case, we cannot simply go on ignoring the limits of our ability to be fully in control of our own behaviour.

Priming in organizations

We often find people in corporations and institutions who are blind to the interdependence between the brain, its environment and the resulting decisions and behaviours. Somehow, many are led to believe that performance and mental agility should be something isolated from external conditions, streaming solely from inside one's head and heart. As this viewpoint goes, it would be great if the company performs fantastically regardless of everything going on around one individual. We find this approach damaging because, above all, it removes the corporation's and leadership's responsibility in creating the right environment for achievement. For us, the individual is always important as far as both the individual and those in the immediate environment are aware and effectively using the key determinants of personal and social behaviour. As the psychologist Philip Zimbardo (2008), who conducted the famous Stanford Prison Experiment mentioned in the previous chapter, often explains, it is not the individual who should always be accused of a behaviour ('the bad apple' approach), nor even the immediate environment ('the barrel of apples' approach), but the ones that have the ability to design the environmental conditions where actors behave and actions take place ('the barrel maker' approach). True leaders are 'barrel makers', creating positive and dynamic environments that prime their own and their employees' behaviour for strong and lasting performance. Below are proven ways of creating the right priming conditions:

- *Culture.* Corporate culture, or even departmental or team culture, is a strong primer for decision making and behaviour. Cultural constructs are found to be key drivers of social behaviour since, when triggering the implicit knowledge of its values, people are automatically put in a specific frame of mind and make behavioural choices accordingly (Hong *et al*, 2000). This means that organizational culture with its actual working values and as an everyday-lived, collective experience – not as a polished presentation or online entry on the corporate website – determines subliminally the way that people's brains are readied for action. It is imperative that you unveil the actual work ethos in your organization and understand its impact on priming people for specific behaviours in every situation, such as in meetings, negotiations, presentations, etc.

- *Context of messaging.* Messages are not independent of the context or media they appear in. Actually, our brains are primed to accept, reject or generally decode and assign meaning to a message based on the implicit perception of the medium, this being a person, an electronic newsletter, or Facebook announcement. This contextual cueing or priming of media (Dahlén, 2005) involves implicit activation of memory and forms attitudes towards one stimulus (the message) based on another (the context or media that the message appears in). This means that meetings, announcements, office walls, the internal portal, informal corridor discussions and different people can have a different priming effect. Choose the medium wisely to make sure it is congruent with the message and not preparing the brain for the opposite reaction.

- *Symbols.* As Alter (2013) explains, symbols can powerfully shape our thinking and acting because we perceive them effortlessly and rapidly, and because they embed themselves deep in our memories. They are 'magnets of meaning', instantly retrieving associations hardwired in our brains by previous experiences and priming us for specific attitudes, decisions and behaviours. Symbols are culture transmitters (Buchanan and Hunczynsky, 2010). In a much-discussed study, researchers (Fitzsimons *et al*, 2008) primed students by exposing them subliminally to either various Apple logos or various IBM logos. Then they asked them to show creativity by suggesting different uses of mundane everyday items, such as a paperclip (the unusual uses test). Those primed with the Apple logos produced more in number and more creative solutions than the IBM group because apparently Apple is wired in the brain as a representation of creative thinking. So, the Apple logo triggers an unconscious reaction that prepares/primes the brain for creativity more than the Microsoft logo. Such findings strongly suggest that the stimuli we expose ourselves and our colleagues to at work can have a potentially powerful knock-on effect on our behaviour. Take a look around you in the office and at the corporate materials you and your team are using daily. Are there any specific symbols that actually boost your brain's performance? Or are there symbols that can inhibit it? Eliminate the latter and enhance the former. Also explore which ones can be introduced and how in order to increase the positive impact of symbols on your team's brain.

- *Words.* Priming has first been observed in exercises with associative words and since then it has been confirmed that it reduces neural processing in the cerebral cortex (Wig *et al*, 2005), meaning that there is less neural activity in the controlled part of our brain than when priming is absent. This is because priming triggers implicit memories that help us perform, mostly automatically, a task based on previous experience without being aware or conscious of this process (Schacter, 1987). Priming saves brain energy and helps us complete tasks efficiently and manage time effectively. Words have increased power in priming and thus we need to pay attention to how we are using them both for our own implicit motivation purposes and those of others in our organizations. In this direction, it makes sense that companies use specifically crafted words for their values that aim to reflect their core philosophy. TOMS Shoes, for example, repeat the word 'give' both in their values and in their corporate materials (TOMS, 2015) to indicate that they are in the business of 'giving' back to communities in need something for every purchase of one of their products (Chu, 2013). There are many lists available of recommended leadership terminology (see for example Brandon, 2015) that can induce positive change and improved behaviour but we suggest that you create your own that fits your role, your aspirations, your organization and your team best. Words will put people, including you, in a specific frame of mind so choose them carefully and strategically.

- *Silence*. More often than not, we have observed a considerable distance between the thinking and the doing worlds in organizations. Office workers are in many cases disconnected with the 'real-world' workers such as factory-level or front-line staff. This kind of disconnect and its risks is what inspired many managers to engage in what is known as management by wandering around (MBWA), a practice aiming at helping those in office to perform reality checks and to ensure improvements, especially when actively engaging in problem solving (Tucker and Singer, 2015). Still, we are often stunned by the amount of talk in offices, in the form of meetings, presentations, reports, etc and the absence of decisive action. An intriguing study by Flegal and Anderson (2008) and the verbal overshadowing effect might explain the reasons for that. They asked highly skilled golfers to first perform an action, then engage in verbal discussions (some explaining their technique, some in irrelevant topics) and at the end to perform the same move again. Describing their experience impaired significantly the more highly skilled golfers' ability to re-achieve results. In contrast, highly skilled golfers who engaged in the irrelevant verbal activity did not damage their performance. This proves that the verbal overshadowing effect can actually harm your and your teams' efforts to repeat great achievements. Too much talk and over-thinking seems to take away from experts their brain's ability to perform. So, fewer endless explanations and more targeted actions pave the way for better brain preparation for repeating great results.

> **Action box: create a priming matrix for priming conditions**
>
> Create a matrix table with six rows and two columns. In the first cell of the first column put the title: aspects of priming conditions. In the next five cells of the first column put one by one the five aforementioned ways that help creating the right priming conditions, namely culture, context of messaging, symbols, words and silence. In the first cell of the second column put the title: organizational elements. Then your target is to fill in the remaining five cells of the second column with those particular aspects that take place in your organization, department or team and whether they could influence in your opinion priming conditions bringing positive or/and negative results.
>
> Write down your conclusions and consider them the next time that you need to take decisions within your team.

New habits, old habits

We might not be aware of it, naturally, but unthinking routines or habits represent a large number of our everyday behaviours. Studies suggest that as much as 45 per cent of our daily actions are repeated in the same location almost every day (Neal *et al*, 2006). This means that almost half of our behaviour is automated and performed with reduced cognitive interference from our side. This mechanism saves vital brain energy channelled to brain areas and functions more needed; since if we had to consciously consider and analyse every action we took every single minute of the day then we would be paralyzed by information overload and we would not be able to achieve much. If a habit is beneficial to our leadership potential then we should keep it and even reinforce it but if it is damaging then we need to reformulate it into something new. Both strategies though assume that we know what habits are and how to deal with them since their massive impact on our behaviour simply cannot be disregarded.

Our neural habit circuits, the brain's regions and connections responsible for habit formation and preservation, have only recently started to be understood. *Reinforcement contingencies* performed by the unconscious mind determine which behaviour will reward us somehow, and which will not, and evaluates the outcome of the chosen behaviour by considering 'reward-prediction error signals' in order to push the formation into a habit or not (Schultz and Romo, 1990). Unconsciously, our brain will nudge us towards a behaviour that it perceives as more rewarding and if this reward actually comes, and it's repeated after we do the same action again and again, then a habit is formed by reinforcement. This is why we might consciously perceive our behaviour as harmful, such as eating junk food too often, but we find it so difficult to break the habit. It is the powerful unconscious mind that decides and evaluates the rewards from its own perspective and pushes for a behaviour, not our weaker controlled thinking. When the unconscious mind loves a reward that we consciously – theoretically – believe can be negative, guess who is usually winning. In the brain, the neocortex, basal ganglia, which directs procedural learning, and the dopamine-emitting midbrain, are associated with habit formation and maintenance, which become more or less engaged as a behaviour is deliberate or habitual (Graybiel and Smith, 2014). In the three-step process of forming a habit, the neocortex is more active in the first step, exploration, when detecting and weighing a new behaviour. In the second step, habit formation, the basal ganglia becomes very active as we repeat a behaviour and the feedback loop is reinforced by the reward of the behaviour. In the third step, imprinting, the habit is well-settled through neuroplasticity and, surprising enough, a part of the neocortex is still active, as if quietly allowing us to engage in the habit (Graybiel and Smith, 2014). And this is the point where hope lies in changing habits.

Formulating a new habit takes time. In a real-world setting and by using human beings (and not mice as in many habit-related experiments), it was found that people formed a habit, which means that automaticity settled in, on average in 66 days with a wide range of 18 to 254 days (Lally et al, 2010). This means that habits do not form overnight and that for neuroplasticity to engrave new settled pathways in the brain, repetition and confirmation of the reward over time are important. Once a habit is formed, new and old goals, expressed by our controlled thinking (conscious), do not interfere much with the process, but for starting new habits goals are important both for starting the process and for maintaining it over time for the habit to be fully formulated (Wood and Neal, 2007).

Similar to individual habits, there are group habits that can be formulated within a collective entity. In management theory, the relevant concept is of *organizational routines* that are seen as a systematic and repetitive set of activities that occur inside organizations (Feldman, 2000) and that, according to some, are the necessary means of task implementation (Pentland and Rueter, 1994). Routines are organizational entities that produce stability, as well as change whenever they are new, within organizations (Pentland et al, 2012). Leaders should consider organizational routines as potentially powerful collective habits that can play a vital role in sustaining and/or changing the status quo within companies.

If you want yourself and your team to develop and sustain new productive habits at the workplace then fully agreeing upon the goal of the habit and ensuring that this goal is emotionally reinforced over time are key steps. For example, if you want to start keeping reflective notes with must-do suggestions after every important meeting but you just don't seem to actually be doing it, setting a very strong goal for it and reminding yourself of this goal every time you enter and exit a meeting can help forming this new habit. Setting up an automatic reminder to pop up in your e-calendar after meetings, explicitly including the goal not just the reminded action, has worked well for our clients. Periodically noting down the improvements you managed to do because of this new habit helps cement it even further through reward reinforcement.

The habit loop is best described by Pulitzer-prize winning reporter Charles Duhigg (2013) in his book *The Power of Habit: Why we do what we do and how to change*. There, he explains the importance of the habit cycle, which goes like this:

1 Cue: The trigger that puts our brain in an automatic mode and asks for a specific habit to be retrieved.
2 Routine: The physical (doing), mental (thinking) and emotional (feeling) process that is triggered by the cue.
3 Reward: The outcome of the routine that the brain craves and the habit exists for.

In order to change an unwanted habit, Duhigg (2013) suggests keeping the cue and rewards the same but choosing a different routine. This is what he calls *the golden rule of habit change*. The essence of this approach is that you can change a habit by trying to achieve the same reward after a specific cue by a different course of action. For example, if you have the habit of automatically accepting more work than your team members in order to achieve the emotional reward of being a 'worthy team member', maybe it is better to help the team deal more efficiently with the new workload by facilitating real teamwork and thus achieving the same emotional reward at the end.

Changing the routine in order to reach the same reward is one way of dealing with habits. Working on cues is another (Wood and Neal, 2007). The first strategy is to avoid the cue altogether. If you do not want to be distracted by impulsively looking at new emails or new phone calls, eliminate the cues by turning your phone to silent or switching off email notifications.

Similarly, if meetings go automatically sour every time a specific team member teases colleagues negatively, talking personally to this team member to stop providing this cue for habitual collective fighting can stop the undesired behaviour. Whenever cues are environment-specific, changing the environment will cause cues to disappear and thus not trigger the habit anymore. Changing meeting locations, for example, and holding meetings in diverse places within and outside the company has worked miracles in our experience for cutting the habit of valueless prolonged discussions that characterize many business meetings. Changing offices, locations, organizational positions and even jobs can work very effectively in the direction of getting rid of unwanted habits. If the cue cannot be eliminated, such as a boss being over-critical in performance evaluation that can cause a habitual submissive response, then being aware of the cue and the routine can be the start of changing the habit. Inhibiting the cue's automatic response involves the strategy of avoidance, which is about vigilantly monitoring cues and their responses in order to change the course of action. This is a difficult strategy because it involves our conscious mind working hard to change behaviour. It works well if monitoring is really vigilant and assisted by peers. Asking those close to us to help keep an eye on our automated habitual responses and to notify us when a cue appears is crucial for our brains to start moving in a different direction. Training, especially as a structured change intervention, is another form of reconditioning the brain to respond differently to environmental cues, as is punishment that can take away the pleasure of fulfilling a habit loop.

Habits are major brain automations and need to be understood and utilized appropriately if we are to succeed as managers and leaders. We need to get rid of habits that do not serve our purpose and to develop new ones that do. Desirable habits save valuable brain energy and time, and help us perform effortlessly with greatness. The global popularity of the book *The Seven Habits of Highly Effective People* by Stephen Covey, first published in 1989 and one of the best-selling business books of all time (Gandel, 2011), demonstrates that people in organizations crave to develop habits that will make them automatically, instinctively, impulsively win. He suggested the following leadership habits: be proactive; begin with the end in mind; put first things first; think win-win; seek first to understand and then to be understood; synergize; and sharpen the saw. The neuroscience behind habit formation and maintenance can help us do so but only if we have a deeper understanding of how and why they work. Which leadership habits do you want to develop? Start today because it will take time. In a true habitual fashion, rewards will be worth the effort.

Action box: create a habit map with the SRHI tool

SRHI, or the Self-Report Habit Index, is a tool developed by Verplanken and Orbell (2003) based on measuring key habit features, such as history of repetitions, automaticity and expression identity of the habit. It involves 12 questions to identify strength of habits. The questions are: Behaviour X is something that...

1 I do frequently.

2 I do automatically.

3 I do without having to consciously remember.

4 Makes me feel weird if I do not do it.

5 I do without thinking.

6 Would require effort not to do.

7 Belongs to my (daily, weekly, monthly) routine.

8 I start doing before I realize I'm doing it.

9 I would find it hard not to do.

10 I have no need to think about doing.

11 Is typically 'me'.

12 I have been doing for a long time.

By scoring each habit on those 12 questions by using an agree/disagree scale of 1 to 5 and calculating the average total score, we can find the overall strength of the habit. We use the SRHI to help people and teams in organizations develop a habit map. This is done by following four main steps:

- Step 1. Identify three good habits you want to retain and three bad ones you want to change. We recommend you ask your closest colleagues in the organization to help you with this.

- Step 2. Complete the SRHI for all habits listed. Ask at least two more people closest to you at work to do the same for your habits, not for theirs. This will give a more independent view on the presence of your habits. Calculate the average scores by combining all responses. Put both the good habits and the bad ones in order of strength (based on their scores), separately. You now have the basis for the map ready.

- Step 3. For each habit, good and bad (and even for completely new ones you want to add to the list), create a list of actions to deal with them effectively. Good habits might need solidifying and even further enhancement while bad ones will need toning down and eventually changing. Follow the recommendations in this part of the chapter for doing so.

- Step 4. Commit to the process by vigilant monitoring, often team discussion, positive and negative reinforcement whenever possible, and adding/subtracting actions as needed. Habits are sticky and cannot be left to develop or change by themselves.

Have in mind that the 12 questions on the SRHI can be altered to reflect more accurately the specific organizational environment and the nature of your work. Also, bear in mind that changing is easier when there is peer support than when it is an isolated, individual effort.

You can do a team habit map by repeating the exercise above but together with all members of your team. You can then calculate team scores and compare them with personal ones.

In both personal and team maps, repeat the exercise six months later at the earliest to identify any improvements. If you realize improvements, then celebrate them!

Let's get physical

The world around us is becoming increasingly more digital. We spend a big part of our days and nights hooked on the internet both for professional and personal reasons. Consequently, this intensive interaction with the digital world does not leave the brain unchanged. On the contrary, our brain is trying to adapt to the new situation as it always does when there are significant changes in our external environment. This brain adaptation is, according to some, positive and, according to others, negative. On the positive side, people like Clive Thompson, author of the book *Smarter than you Think: How technology is changing our minds for the better* (2013), believe that our brain becomes more efficient and effective with the use of new technologies because of advanced collaboration opportunities, easier access to critical information globally and the ability keep track of every aspect of our lives. On the negative side, people like Larry Rosen, author of the book *iDisorder:*

Understanding our obsession with technology and overcoming its hold on us (2012), believe that our obsession with mobile devices has already shown signs of pathological brain conditions, such as addiction, heightened narcissism, attention deficit and obsessive-compulsive behaviours. These are obviously different sides of the same coin and it remains to be seen how our brains will benefit, but also suffer, in the long run from these technologies. The fact is though that our brains have evolved over millennia through intensive interaction with the physical world and this interaction has created numerous rule-of-thumb types of thinking and reacting in our unconscious mind. The stimuli we expose our senses to daily are signalling to our brain in much more strength than we thought of, influencing our unconscious mind to behave in surprising ways. Being aware and understanding those real-world influences on our powerful concealed thinking can put us in the advantageous position of manipulating them according to our goals. Welcome to the world of 'matter over mind'.

We are our brains, situated within our physical bodies and living in a specific environment that surrounds us constantly. Isolating the one from the other simply does not make sense. The concept of embodied cognition in philosophy and psychology argues convincingly that we cannot separate our brain from its body and the body from its environment. The interrelation of brain, body and the wider biological, psychological and social environment we find ourselves in is critical in how our mind works and it directly impacts the way we think, feel and behave (Valera *et al*, 1991). According to this approach, most of our cognition is caused from the interplay of our *sensorimotor* brain regions, which is the neural integration of our sensory and motor systems that help us, receive stimuli and move accordingly, with our bodies and the environment we interact with. Evidently, scientists developing artificial intelligence (AI) systems have long recognized that the problem with AI and robotics is not replicating high-level reasoning that actually requires less computation. The challenge is to replicate lower-level *sensorimotor* abilities that need huge amounts of mostly unconscious computations and that are the ones responsible for interacting effectively and efficiently with the world around us in order to survive and thrive (Moravec, 1988). The power of the unconscious is what makes us unique as a species and we can no longer pretend that it does not really matter or that it mostly damages our supreme intellect. We should know better. Leaders need to know better if they are to improve their teams, their organizations and their societies.

Our mind's dependence on the physical world engulfing it is also vividly reflected in spoken language since we constantly use real-world metaphors to express ourselves on an everyday basis (McGilchrist, 2012). So, we have a 'heavy' schedule this week, a 'clear' decision was made yesterday, we constantly 'push' for reforms, our opponents finally 'dropped' the charges against our firm, the team did not really 'grasp' the importance of the latest announcement, and the new CEO is rather 'cold' compared to the previous one. The list is endless showing that through evolution our brains have developed a profound ability to deeply consider the natural world around it, even sometimes without separating it totally itself from it, in order to always have a realistic view of the situation and react in the best possible way. The fact that a big part of modern thought has been based on the premise that we can somehow be objective and detached observers of the events that surround us, is considered by many psychologists, neuroscientists and philosophers alike as not just false but as highly dangerous (McGilchrist, 2012). If we do not interact constantly and actively with the world around us, utilizing all our senses and powerful brain functions, we are in danger of ending up 'living up there, in our heads', isolated from reality, as Sir Ken Robinson accused many academics and modern professionals doing, in the most-watched TED speech of all time (Robinson, 2006). And this is, unfortunately, what we personally experience in many companies and training sessions we are involved with. There are many people in organizations globally who have been wrongly led to believe that the mind has a life of its own, separate from the brain, the body and the environment, and that it presides over all of them. They are too analytical, too dry, too disconnected and they firmly believe that this is the best option for a true leader. This cannot be further from how our brains actually work and what true leaders can do to inspire, perform and grow.

How important is the actual physical environment to our behaviour? Consider the following experimental research results:

- Holding briefly a cup of hot beverage instead of an iced one makes us judge people across from us as having 'warmer, more generous, more caring' personalities (Williams and Bargh, 2008).

- Similarly, if we are asked to hold a hot therapeutic pad instead of a cold one, we are more likely to select a gift for a friend than for ourselves (Williams and Bargh, 2008).

- If we are holding a heavy clipboard with information instead of a lighter one, we assign higher monetary values to foreign currencies, increase fairness in our decision making, are more consistent in our deliberations and tend to hold stronger opinions (Jostmann *et al*, 2009).

- In a series of experiments Ackerman and associates (2010) found that sitting on a hard chair will make us negotiate harder, solving a hardened puzzle will make us less socially cooperative after the game, and examining by touch a soft blanket will make us consider someone's personality as more positive than when examining a hard object.

Highlighting the influence of haptic (or touch) sensations for our whole existence, Ackerman *et al* (2010) argue that:

> Touch is both the first sense to develop and a critical means of information acquisition and environmental manipulation. Physical touch experiences may create an ontological scaffold for the development of intrapersonal and interpersonal conceptual and metaphorical knowledge, as well as a springboard for the application of this knowledge.

Whenever we present such scientific evidence to our audiences and clients they usually react with great surprise and awe, and a few with rigid scepticism. It is difficult to accept that holding a hot cup of coffee or sitting on a hard chair can really change organizational behaviour within companies and institutions. This shows how much we have removed ourselves mentally from our physical environment, considering our cognitive processes as independent, unaffected and objective. The amount of such research results that indicate the opposite is increasing and has led psychologist Thalma Lobel (2014) to write the book *Sensation: The new science of physical intelligence*, the latter being another name for embodied cognition in the cases when it works to our benefit and not to our confusion and misguided behaviour. Physical intelligence is actually an excellent term indicating that our brains' ability to take cues from the physical environment in order to move in one direction or another can provide important competitive advantages to those who know how to spot this ability and utilize it. Lobel has included textures, heaviness, colours, distance, brightness, cleanness, highs and lows, as well as the metaphor-based language we use daily, in her analysis of how our senses influence our unconscious mind and prime us for action. She concludes that:

> Once you become aware of these influences and the power of metaphors, you can use them for your benefit... Pay attention to your senses' input and evaluate it. Your attunement to what your senses tell you will give you physical intelligence – otherwise your senses will yield only data. Armed with this new awareness, you might avoid being swayed by previously unconscious metaphorical associations in your judgments and evaluations of others.

This is the invaluable lesson for embodied cognition; ignoring or denying its existence actually increases the possibility of operating with biases and having a distorted view of reality. This simply means that we can take bad decisions and misguided actions. Accepting it and working with it can increase our effectiveness and ultimately it can become a powerful weapon in your brain-based leadership style. Our recommendation is always for our partners to start feeling the environment around them in a concrete and clearer way. Noticing details of the space around them and searching deep inside to feel how these might change the way they think and act is a good start. We particularly ask them to search for how the physical environment affects:

A Their everyday mood.

B Repetition of specific chains of thought.

C Closeness and connectedness levels to colleagues and other people in the organization.

D Their creative potential and innovative reflexes.

Even how we form our body postures is priming our emotional state in surprisingly powerful ways. We can literally fake a posture to produce the desired mental effect, which means our bodies can literally change our minds (Cuddy, 2012). Remember the mouth-holding pencil exercise that produces happy feelings just by creating a fake smile? The interaction between our unconscious brain, our bodies and the physical environment around us creates powerful priming effects to our conscious mind that if not acknowledged and managed can lead to considerable mistakes in both thinking and doing. Having said that, evolution has taught this system what works and what is not working so listening to our gut feeling does not mean automatically rejecting it. Actually, gut feelings and intuition play the most significant role in developing leadership expertise for complex and dynamic environments.

Expertise and automaticity

The more we think about a task analytically, chances are that the less expert we are in this task. Think about it. While we learn to ride a bike we need to think about all possible aspects of this task in order to learn how to do it and avoid breaking an arm. But when we do learn it, there is not much thinking going on anymore while riding the bike. We can literally ride the bike in autopilot and think about something else at the same time. The more you think about a problem, a task or a new idea the more it shows that expertise has to be developed and not that it is already present. In a seminal study of chess master players, psychologist and chess expert Adriaan de Groot and his associate (1996) found that chess masters, the ultimate experts of the game, have a remarkable ability 'to quickly capture, and to retain and recall, the information contained in an unknown, complex chess position: its gist and its structure, up to the precise location of almost all men'. They found that experts quickly processed new information and produced an intuitive reaction (through a heuristic or shortcut in the brain) faster than they produced a controlled, conscious thought. Expertise, although following complicated pathways in our brains, largely lives in a subcortical structure called cerebellum, or 'the little brain'. This part of the brain has been recently linked to the evolution of our unique human characteristics, such as technical intelligence, since it developed much faster than the neocortex in our ancestors (Barton and Venditti, 2014). It has even been surprisingly linked to our creative abilities (Saggar *et al*, 2015). These latest scientific insights show how important it is for our thinking and acting in our everyday working lives. Although older, cortico-centric approaches considered expertise and 'gifted' personal characteristics as a top-down process within the brain, new research suggests that the cerebellum plays a huge role in expertise development since this is where behavioural refinement takes place (Koziol *et al*, 2010). Traditionally, the cerebellum was thought to be doing this refinement of final action only in relation to motor-related behaviours, but more recently this has been extended to other brain functions involving learning (Doya, 2010). Simply put, 'practice makes perfect' is a cerebellum responsibility and it does not live in the conscious mind. You have to go deeper to find real expertise.

This does not mean though that all expertise is fully automatic. On the contrary, thinking and behaving in critical situations when new solutions are needed urgently is a collaborative effort of the conscious and the unconscious minds. Expertise is fine when its fast pattern recognition is applied to rather routine tasks. But when we face chaotic and multifaceted situations, automatic pattern recognition has to be accompanied by fast mental simulations of a few possible outcomes to bring the best results. Pattern recognition alone is highly risky in challenging times, as we saw in Chapter 2. This is the Recognition-Primed Decision (RPD) model developed by Gary Klein (1998), initially based on interviews of firefighters' commanders in the United States, as a branch of naturalistic decision making which is mostly intuitive and automatic. In crisis situations such as a blazing fire, pattern recognition is very helpful, but could bring the wrong decision because of rapid-changing new conditions, and deliberate analysis alone could be too slow to offer any value at all. So, intuitively retrieving patterns from your long-term memory that you can trust, but at the same time performing a fast and mini scenario analysis (up to three scenarios) in your conscious mind based on the conditions you see, can be the recipe for leadership expertise in our times.

Our experience shows that influential and inspirational leaders in organizations know how to use their experience as an important decision-making factor, but always contrast this experience and their intuition with current conditions, other people's input and swift scenario analysis. They trust their gut, especially when it sends strong signals towards a decision or a behaviour, but they know better to process it mentally to see where it can lead compared to other possible alternatives. For us, this is clearly the beauty of our brains: nothing happens in isolation and everything is related. This is why we need to try to engage the right parts for the right action, and that is a core aspect of our brain adaptive leadership model.

> ### Boost your brain
>
> Try to remember and write down a minimum of three and maximum of five moments from the last year where you demonstrated to your peers a high level of expertise. This could be a vital contribution during a difficult meeting, an answer to a challenging question after a presentation, a winning point in tough negotiations etc. The focus should be on identifying those heated moments where your expertise proved invaluable for instant problem solving and when your colleagues were clearly impressed by this demonstration of advanced expertise. For each of these moments create a table with the attributes of your expert intervention. Where did it come from? Experience, intuition, fast pattern recognition, conscious analysis? How much did each of these play a role and how? By unveiling the hidden forces of expertise in dynamic situations you will be able to understand them better and utilize them more effectively next time you need to save the day in your organization.
>
> Continue this exercise in the future whenever you recognize such a moment again because this will allow you to include more expert moments and thus help you build a long-term understanding of your expertise and how to improve it.

Keep in mind

If you separate the mind from the brain and selectively marvel at the supposed supremacy of conscious thinking over intuition and gut feelings you will constantly be disappointed in yourself and often irritated by those around you. This is because we cannot think and act consciously most of the time and even more, it is not neurologically possible to 'free' our minds and therefore our decisions and actions from gut feelings and intuition. Actually, it would be dangerous to do so since those provide necessary deep-wired intelligence much needed in our everyday lives as well as free space for our conscious mind to focus on other matters. Adopting your brain's automaticity processes as key elements of what can make you a great leader can drastically improve both your understanding of yourself and others, and remarkably enhance your performance. Using priming to positively and appropriately prepare your brain and the brains in your team, establishing which work habits you need to abolish and which to promote and creating a physical environment that nudges you and everybody else in the right direction are crucial for mastering the brain's automatic responses. Becoming a true expert in challenging times requires you to combine this knowledge of how heuristics, intuitions and habits work together with fast analytical, clear and targeted thinking. Taking into account the unconscious mind's power over our thinking and doing, it all starts from here: what is the readiness potential of your leadership brain?

References

Ackerman, JM, Nocera, CC and Bargh, JA (2010) Incidental haptic sensations influence social judgments and decisions, *Science*, **328**, pp 1712–1715

Alter, A (2013) *Drunk Tank Pink: The subconscious forces that shape how we think, feel, and behave*, Oneworld Publications, London

Bargh, JA (1994) The four horsemen of automaticity: Awareness, efficiency, intention, and control in social cognition, in *Handbook of Social Cognition*, 2nd edn, eds RS Wyer and TK Srull, pp 1–40, Erlbaum, Hillsdale

Bargh, JA (2013) Social psychology cares about causal conscious thought, not free will per se, *British Journal of Social Psychology*, **52** (2), pp 228–230

Bargh, JA (2014) Our unconscious mind, *Scientific American*, January, **310** (1), pp 20–27

Bargh, JA, Chen, M and Burrows, L (1996) Automaticity of social behavior: Direct effects of trait construct and stereotype activation on action, *Journal of Personality and Social Psychology*, **71** (2), pp 230–244

Barton, RA and Venditti, C (2014) Rapid evolution of the cerebellum in humans and other great apes, *Current Biology*, **24** (20), pp 2440–2444

Bos, MW, Dijksterhuis, A and van Baaren, RB (2006) On making the right choice: The deliberation-without-attention effect, *Science*, 17 February, **311**, pp 1005–1007

Brandon, J (2015) 40 powerful words to help you lead a team, *Inc.*, URL: www.inc.com/john-brandon/40-words-of-leadership-wisdom.html, accessed 20 July 2015

Buchanan, DA and Hunczynsky, AA (2010) *Organizational Behaviour*, 7th edn, Prentice Hall, London

Cambridge Advanced Learner's Dictionary and Thesaurus © Cambridge University Press, URL: http://dictionary.cambridge.org/dictionary/english/unconscious

Chu, J (2013) TOMS sets out to sell a lifestyle, not just shoes, *Fast Company*, URL: www.fastcompany.com/3012568/blake-mycoskie-toms, accessed 26 July 2015

Cuddy, A (2012) Your body language shapes who you are, *TED Global*, URL: www.ted.com/talks/amy_cuddy_your_body_language_shapes_who_you_are?language=en, accessed 10 March 2015

Dahlén, M (2005) The medium as a contextual cue: Effects of creative media choice, *Journal of Advertising*, **34** (3), pp 89–98

Doya, K (2000) Complementary roles of basal ganglia and cerebellum in learning and motor control, *Current Opinion in Neurobiology*, **10** (6), pp 732–739

Doyen, S, Klein, O, Pichon, CL and Cleeremans, A (2012) Behavioral priming: It's all in the mind, but whose mind? *PLoS ONE*, **7** (1), e29081

Duhigg, C (2013) *The Power of Habit: Why we do what we do and how to change*, Random House, London

Fedor, DB, Davis, WD, Maslyn, JM and Mathieson, K (2001) Performance improvement efforts in response to negative feedback: The role of source power and recipient self-esteem, *Journal of Management*, **27**, pp 79–97

Feldman, M (2000) Organisational routines as a source of continuous change, *Organization Science*, **1**, pp 611–629.

Fitzsimmons, GM, Chartrand, TL and Fitzsimmons, GJ (2008) Automatic effects of brand exposure on motivated behavior: How Apple makes you 'think different', *Journal of Consumer Research*, **35** (1), pp 21–35

Flegal, KE and Anderson, MC (2008) Overthinking skilled motor performance: Or why those who teach can't do, *Psychonomic Bulletin and Review*, **15**, pp 927–932

Gandel, S (2011) The 7 habits of highly effective people (1989), by Stephen R. Covey in the 25 most influential business management books, *Time*, 9 August, URL: http://content.time.com/time/specials/packages/article/0,28804,2086680_2086683_2087685,00.html, accessed 5 May 2015.

Gazzaniga, MS (1998) *The Mind's Past*, University of California Press, Berkeley

Gigerenzer, G (2007) *Gut Feelings: Short cuts to better decision making*, Penguin, London

Gladwell, M (2005) *Blink: The power of thinking without thinking*, Penguin Books, London

Groot, AD de and Gobet, F (1996) *Perception and Memory in Chess: Studies in the heuristics of the professional eye*, Van Gorcum, Assen

Hong, YY, Morris, MW, Chiu, CY and Benet-Martinez, V (2000) Multicultural minds: A dynamic constructivist approach to culture and cognition, *American Psychologist*, **55** (7), pp 709–720

Jostmann, N, Lakens, D and Schubert, T (2009) Weight as an embodiment of importance, *Psychological Science*, September, DOI: 10.1111/j.1467-9280.2009.02426.x.

Kahneman, D (2012) A proposal to deal with questions about priming effects, *Nature* online, URL: www.nature.com/polopoly_fs/7.6716.1349271308!/suppinfoFile/Kahneman%20Letter.pdf, accessed 10 June 2015

Klein, GA (1998) *Sources of Power: How people make decisions*, MIT Press, Cambridge, MA

Koziol, LF, Budding, DE and Chidekel, D (2010) Adaptation, expertise, and giftedness: Towards an understanding of cortical, subcortical, and cerebellar network contributions, *The Cerebellum*, **9** (4), pp 499–529

Kristoff, ND (2008) What? Me biased?, *The New York Times* online, 29 October, URL: www.nytimes.com/2008/10/30/opinion/30kristof.html?_r=2, accessed 5 April 2015

Lally, P, Van Jaarsveld, CH, Potts, HW and Wardle, J (2010) How are habits formed: Modeling habit formation in the real world, *European Journal of Social Psychology*, **40** (6), pp 998–1009

Libet, B, Gleason, CA, Wright, EW and Pearl, DK (1983) Time of conscious intention to act in relation to onset of cerebral activity (readiness-potential) – the unconscious initiation of a freely voluntary act, *Brain*, **106** (3), pp 623–642

Lobel, T (2014) *Sensation: The new science of physical intelligence*, Icon Books, London

McGilchrist, I (2012) *The Master and His Emissary: The divided brain and the making of the Western world*, Yale University Press, New Haven

McGowan, K (2014) The Second Coming of Sigmund Freud, *Discover*, 24 April, pp 54–61

Meyer, DE and Schvaneveldt, RW (1971) Facilitation in recognizing pairs of words: Evidence of a dependence between retrieval operations, *Journal of Experimental Psychology*, **90**, pp 227–234

Moravec, H (1988) *Mind Children: The future of robot and human intelligence*, Harvard University Press, Cambridge

Neal, DT, Wood, W and Quinn, JM (2006) Habits – a repeat performance, *Current Directions in Psychological Science*, **15** (4), pp 198–202

Newell, BR, Wong, KY, Cheung, JC and Rakow, T (2009) Think, blink or sleep on it? The impact of modes of thought on complex decision making, *The Quarterly Journal of Experimental Psychology*, **62** (4), pp 707–732

Pentland, B and Rueter, H (1994) Organizational routines as grammars of action, *Administrative Science Quarterly*, **39**, pp 484–510

Pentland, BT, Feldman, MS, Becker, MC and Liu, P (2012) Dynamics of organisational routines: A generative model, *Journal of Management Studies*, **49**, pp 1484–1508

Robinson, K (2006) Ken Robinson: Do schools kill creativity? [video file], *TED*, URL: www.ted.com/talks/ken_robinson_says_schools_kill_creativity?language=en, accessed 15 May 2015

Rosen, LD (2012) *iDisorder: Understanding our obsession with technology and overcoming its hold on us*, Macmillan, New York

Saggar, M, Quintin, EM, Kienitz, E, Bott, NT, Sun, Z, Hong, WC, Chien NL, Dougherty, RF, Royalty, A, Hawthrone, G and Reiss, AL (2015) Pictionary-based fMRI paradigm to study the neural correlates of spontaneous improvisation and figural creativity, *Scientific Reports*, **5**, article 10864

Schacter, DL (1987) Implicit memory: History and current status, *Journal of Experimental Psychology: Learning, Memory, and Cognition*, **13**, pp 501–518

Schultz, W and Romo, R (1990) Dopamine neurons of the monkey midbrain: Contingencies of responses to stimuli eliciting immediate behavioral reactions, *Journal of Neurophysiology*, **63** (3), pp 607–624

Soon, CS, Brass, M, Heinze, HJ and Haynes, JD (2008) Unconscious determinants of free decisions in the human brain, *Nature Neuroscience*, **11** (5), pp 543–545

Thompson, C (2013) *Smarter than you Think: How technology is changing our minds for the better*, William Collins, London

TOMS (2015) Improving lives, *TOMS online*, URL: www.toms.com/improving-lives, accessed 26 July 2015

Tucker, AL and Singer, SJ (2015) The effectiveness of management-by-walking-around: A randomized field study, *Production and Operations Management*, **24** (2), pp 253–271

Valera, FJ, Thompson, E and Rosch, E (1991) *The Embodied Mind: Cognitive science and human experience*, MIT Press, Cambridge, MA

Verplanken, B and Orbell, S (2003) Reflections on past behavior: A self-report index of habit strength, *Journal of Applied Social Psychology*, **33** (6), pp 1313–1330

Wig, GS, Grafton, ST, Demos, KE and Kelley, WM (2005) Reductions in neural activity underlie behavioral components of repetition priming, *Nature Neuroscience*, **8** (9), pp 1228–1233

Williams, LE and Bargh, JA (2008) Experiencing physical warmth promotes interpersonal warmth, *Science*, **322**, pp 606–607

Wilson, TD (2002) *Strangers to Ourselves: Discovering the adaptive unconscious*, Harvard University Press, Cambridge, MA

Winerman, L (2011) Suppressing the 'white bears', *APA Monitor on Psychology*, October, **42** (9), p 44

Wood, W and Neal, DT (2007) A new look at habits and the habit-goal interface, *Psychological Review*, **114** (4), pp 843–863

Zaltman, G (2003) *How Consumers Think: Essential insights into the mind of the market*, Harvard Business School Press, Boston

Zimbardo, P (2008) Philip Zimbardo: The Psychology of Evil [video file], *TED*, URL: www.ted.com/talks/philip_zimbardo_on_the_psychology_of_evil?language=en, accessed 20 February 2015

SUMMARY OF PILLAR 3: BRAIN AUTOMATIONS

TABLE S.3 Summary of pillar 3

Understand how priming works	Priming is a process through which we can prepare the brain for specific thoughts and behaviours and it works well when people are not aware of its effect. Five ways of creating the right priming conditions in organizations: • emphasize cultural values; • be aware of the context of messaging; • do not neglect symbols; • choose the wording; • do not underestimate silence.
Maintain or change habits	There is a three-step process of getting and maintaining a habit: 1 Exploration – a new behaviour is detected, weighted and accepted. 2 Habit formation – constantly repeat a behaviour that is reinforced by a reward. 3 Imprinting – the habit is well-settled. You can change unfavourable habits and establish more desired ones by changing the cue and by selecting alternative routines to achieve the same rewards.
Familiarize with the physical environment	The interrelation of brain, body and the wider biological, psychological and social environment is critical in how our mind works and it influences the way we think, feel and behave as leaders. Continuously search how the physical environment affects: • mood; • repetition of specific chains of thought; • closeness and connectedness to colleagues and other people in the organization; • creative potential and innovative reflexes.

PILLAR 4
Relations

More connected, more successful

07

Success, it's a lonely path

She doesn't trust people. It's not that she can't work with them; on the contrary, she has participated in, and even created as a unit director in her current position, effective teams in the various organizations she has worked for. Very early in her career she learned to be cautious, if not suspicious, of her colleagues. She specifically remembers a case where, as a young manager, she was betrayed by another manager she trusted very much. She used to tell him openly about all her doubts and reservations about some of the company's moves and he went on to reveal all these to her director. When she found out about this she felt utterly betrayed and, as recommended by her friends and family, she started trusting no one at work. If she was going to make it to the top, it would be a lonely path.

This approach served her well. As she often says to her friends, it is better to be safe than sorry with relationships at work. 'Always keep yourself to yourself and everything will work out fine' is her motto, and consequently she does not seem to her colleagues very open to warm and strong relationships. While she is approachable, friendly and hands-on to anyone asking for her assistance, as soon as people try to get emotionally any closer they hit a wall. She has boundaries that are known to all. She understands that in some cases she probably meets people that deserve more from her but she is not willing to risk anything. 'Better safe than sorry,' she says. All this, until her deputy director left the company and had an exit interview with her. That discussion was the beginning of revising and eventually changing her strategy towards personal relationships at work.

> Her deputy was much younger than her but managed to climb up the ladder fast, always considered a young star in the company. The fact that she decided to leave was a blow for the trust-impaired unit director since she was planning for the deputy to replace her as she was to move further up the hierarchy. Why did the deputy decide to go? The answer was simple: 'I didn't feel I could get close to you,' she said to the unit director. 'Regardless of the successful cooperation all these years, I still felt I could not trust you fully and become something more than typical supervisor and deputy, but more like your coach and mentor, with whom you can discuss critical issues,' she continued. 'I want something more from the people I work so closely and intensively with,' she concluded. Probing more, the director understood that things were changing at work. Intensive competition both from outside and inside the company, various economic crises around the world, the new hyper socially wired lives we lead, and the unpredictability and dynamism of everyday work at the office, require a different type of relationship. Creative solutions to chaotic and complex new problems ask for closer, deeper and stronger types of relationships within and between teams. The 'I am a lone ranger; I trust no one; I succeed by myself' philosophy that probably worked well in some earlier decades cannot be applied anymore. We work, create, solve and live together at work in fast rhythms never seen before. Success or failure depends more on the speed and effectiveness of everyday collaboration than on our actual, job-related skills and knowledge. This simple but powerful realization shook the director's belief of 'going at it emotionally alone' and through intensive discussions, personal studying and the relevant coaching she has now become a master of relations and an expert human collaborator.

Is the best human strategy for success one of cooperation or conflict? Economists and behavioural scientists tried to answer this question through many theoretical constructs, most notably the famous, or infamous depending on how you look at it, prisoner's dilemma. Although conceived to prove that humans are better off non-collaborating than collaborating, in real life it actually proves the opposite (see page 165 for a detailed explanation). The mindset of a lone ranger or a lone wolf succeeding in all arenas and winning all challenges is an out-dated and very dangerous one. Although there might be cases where our individual contribution is more desired and effective than a collective one, such cases are progressively becoming the exception and not the rule. The good news is that our brain is profoundly wired for collaboration and social bonding, and that by understanding the neurological and behavioural underpinnings of cooperation we can reap the benefits faster and better than ever before. Our brains are inherently very good at it; for collaboration to work its miracles, though, we need to want it as well.

The prisoner's dilemma for humans

The prisoner's dilemma represents for us one of the strongest examples of the clash between over-analytical thinking and natural human behaviour or, as some of our students like to call it, of theory vs reality. Although the prisoner's dilemma was (and still is) used by some economists as a mental exercise to show that selfish behaviour is preferable to cooperation, it proves exactly the opposite! As the renowned economist and Nobel Laureate Amartya Sen (Nobel Memorial Prize in Economics in 1998) has said, 'the purely economic man is indeed close to being a social moron. Economic theory has been much preoccupied with this rational fool' (Thaler, 2015). And this is more so in the case of the prisoner's dilemma.

The game is simple. Two people have just been arrested by the police. They have been separated and offered two options in their interrogation rooms: they can either stay silent (the 'cooperate' strategy) or confess their crime (the 'defect' strategy). The consequences of those choices are the following (Thaler, 2015):

- If they both remain silent (so both choose to cooperate) they are only charged with a minor crime and exit jail in a year.

- If they both confess (so both choose to defect) they each get five years in jail.

- If one confesses and the other doesn't (one defects and one cooperates) then the confessor is free and the other gets 10 years.

What would you do?

We vividly remember the time when, as students, we were first presented with this dilemma. The calculated solution was to defect because the best possible outcome of defecting was better than the best possible of cooperating (free vs one year in jail) and the worse possible outcome of defecting was also better than that of cooperating (five vs 10 years in jail). When one of us voiced this calculated opinion in class the economics professor shouted: 'Excellent! This is a prime example of rational thinking. Bravo!' What a victory for economics thinking. The problem is, of course, that by following this strategy we never get the optimum outcome for both which is the one year in jail for each. But who cares, since mainstream economics wants humans to be selfish, in the same way that Richard Dawkins' (1976) genes are always acting selfishly to survive within future generations. For mainstream economics thinking selfishness is the rational way, while cooperation is the primitive, emotional one.

The celebrated, Nobel Laureate economist John Nash, his life captured in the 2001 Oscar-winning film *A Beautiful Mind* with Russell Crow, mathematically worked out his Nash Unique Equilibrium as a solution to social games such as the prisoner's dilemma, to highlight that the best possible and rational strategy is not to cooperate (Capraro, 2013). However, this is not what happens in reality. Richard H Thaler, one of the fathers of behavioural economics and co-author with CR Sunstein of the 2009 seminal book *Nudge: Improving decisions about health, wealth and happiness*, with a slightly ironic tone towards traditional economic thinkers that often get surprised when encountering real human behaviour, recently reported that:

> *The game rhetoric prediction (for The Prisoner's Dilemma) is that both players will defect, no matter what the other player does, it is in the selfish best interest of each player to do so. Yet, when this game is played in the laboratory, 40–50% of the players cooperate, which means [for traditional economic thinkers] that about half the players either do not understand the logic of the game or [for those more open to accept natural human behavior] feel that cooperating is just the right thing to do, or possibly both. (Thaler, 2015)*

The main problem with the prisoner's dilemma is that, as with many concepts in classical economics, it is separated from real conditions. It assumes that people cannot communicate during the process of making a decision, they cannot change their mind, they do not really know each other and they play the game only once. In reality and especially in the business and managerial situations we face daily, all these conditions are actually in reverse (Dimitriadis and Ketikidis, 1999). In many situations, if not in all within our organizations, we know people and they know us, we will most probably continue interacting with them after a specific instance, we can communicate and engage with them while searching for the best strategy and we have reputations that precede us and that will follow us for a long time. Thus, the best strategy is often to cooperate. Even in actual jail-related situations, cooperation seems to be the default strategy and not the exotic option as economists theorized. In a *WIRED* magazine article about two caught Las Vegas gamblers who managed to illegally hack the algorithm of electronic games in casinos, it was reported than when law officers offered them a version of the prisoner's dilemma in order to push them to confess, they actually remained silent and walked free a few months later (Poulsen, 2014). This, despite the fact they had a very bad argument in the period before, and leading up to, their arrest. Even when the game is played in sterile lab environments many do prefer to cooperate, despite economists' prediction, since we are often inherently optimistic that the best outcome for both players will prevail (Capraro, 2013). The prisoner's dilemma for humans suggests collaboration, for homo economicus, defection. In which species do you and your colleagues belong?

On the practical side of choosing the cooperative option as a sustainable leadership strategy, two books had a profound impact on us: Robert Axelrod's classic *The Evolution of Cooperation* (Axelrod, 1984) and the more popular and accessible *The Origins of Virtue: Human instincts and the evolution of cooperation* by Matt Ridley (1997). Axelrod's main contribution was the fact that he organized a competition for software developers to submit specially developed programs and win a prisoner's dilemma contest. Those software programs competed with each other, simulating prisoners in the prisoner's dilemma game. The programs had to include rules of behaviour based on the action and reaction of the other programs. They were asked to engage with each other in multiple rounds of the game each time. He found that the most successful programs that won repeated games were not the ones favouring the defect option nor the ones blindly cooperating under any circumstances. Based on his findings and suggestions, here are specific strategic steps for modern leaders to maximize the positive effect of cooperation:

- *Be nice: Never defect first, always start by cooperating.* If you defect first, people will take it as your default strategy and they will approach you with extreme caution. In difficult negotiations, in meetings with people you don't know and in interactions with people outside the company, always start by offering your partnership by demonstrating goodwill. In that way you increase the possibility of them cooperating as well.
- *Retaliate/reciprocate: If they defect, defect too.* Do not allow a hostile move to go unnoticed. By defecting when they do, you give a strong message 'I am not a fool; do not take me for granted.' If they cooperate, cooperate too without a second thought.
- *Be forgiving: Defect when they do but after that show goodwill again.* This is the only way to get out of a vicious and eternal circle of defections that do not produce the optimum outcome for all parties involved. Forgiving and demonstrating a will for better future relations is a winning strategic choice.

- *Communicate: Reality-based versions of the dilemma occurring in our everyday office lives do not require people to be isolated in detention cells.* Actively exchanging reliable information can do miracles for achieving commonly beneficial outcomes. Talk to people, observe their verbal and non-verbal responses, and avoid consciously misleading and tricking your counterparts.
- *Be non-envious: Focus on maximizing your own score and not necessarily on achieving the best total score in one round.* Surprisingly, those in Axelrod's tournament trying to be on top in every round ended up losing the top spot overall. Thus, focusing on personal and not absolute progress helps you achieve the best results for all.

We have been applying and teaching those principles for many years now and they have always worked well. We strongly recommend you apply them too, carefully but decisively. You will experience some positive results immediately, but the further you apply them the more long-term the benefits will become.

Matt Ridley (1997) provided a very comprehensive account of the phenomena of cooperation, altruism and morality and offers an excellent introduction to the subject. Trying to combine the selfish gene theory with the widespread phenomenon of collaboration, he claimed that 'selfish genes sometimes use selfless individuals to achieve their ends'. He concludes his work by stressing that 'our minds have been built by selfish genes, but they have been built to be socially trustworthy, and cooperative. This is the paradox this [his] book has tried to explain'. Tit-for-tat, the reciprocal strategy of 'I'll scratch your back if you scratch mine' is central to his analysis of why we collaborate outside our kinship and why these collaborations succeed. And it does not have to come from our analytical mind since reciprocity is '... an instinct. We do not need to reason our way to the conclusion that "one good turn deserves another"' (Ridley, 1997). As best described by Axelrod (1984):

> *There is a lesson in the fact that simple reciprocity succeeds without doing better than anyone with whom it interacts. It succeeds by eliciting cooperation from others, not by defeating them. We are used to thinking about competitions in which there is only one winner, competitions such as football or chess. But the world is rarely like that. In a vast range of situations, mutual cooperation can be better for both sides than mutual defection. The key to doing well lies not in overcoming others, but in eliciting their cooperation.*

> **Action box: the prisoner's dilemma in your team**
>
> Separate your team into two groups. Each group should have at least four people. Take the first group and put its members in different rooms: each person must be alone in each room. Do not tell them who exactly their personal partner in the game is. Present each one separately with the prisoner's dilemma rules, tell them that they will play it only once, and ask them for their decision: will they cooperate or defect? Repeat the exercise with the second group but this time while all being in the same room, face to face with each other and with the ability to play the game three times in a row. Note the decisions of both groups. Discuss with all any possible differences in the strategies between the two groups. Why do those differences exist? Expand the discussion by asking them what the best strategy in the real word usually is. Ask them to refer to real examples from their own experience. You will probably notice that rationality-biased participants will argue hard for the value of defecting while more natural relationship-builders will be prone to favour the cooperation strategy. Try to build a consensus around Axelrod's steps mentioned in the box 'The prisoner's dilemma for humans'.

The socially wired brain

How important are social connections in our lives for who we are, how we behave and what we achieve? Is our brain a personal score-maximization machine or a collective-balancing one? Should we only care about ourselves or should we spend our valuable brain energy on others too? Ultimately, is leadership an individual or a social game? Such questions are central to the BAL model since increasing evidence from neuroscience, psychology and sociology suggest that 'WE' is more important than 'I' in surprising and sometimes counterintuitive ways. A leader who cannot understand and connect meaningfully with the brains of others is a leader doomed to fail.

Our brain is a social organ. Social relationships have a dramatic impact on a large number of brain functions and connections, and ignoring them or downgrading them can only harm our overall leadership performance. The mere fact that we have a consciousness, observing and reflecting on our actions and on the surrounding world, is because of our social dimension. Eminent psychologists and neuroscientists attribute the very existence of consciousness to social relationships. Consciousness, for Professors of Psychology Peter Halligan and David Oakley in the United Kingdom (Halligan and Oakley, 2015):

> ... simply occurs too late to affect the outcomes of the mental processes apparently linked to it... We suggest it is the product of our unconscious mind, and provides an evolutionary advantage that developed for the benefit of the social group, not the individual.

This is an amazing statement. The very thing that seems to make us different from each other and even egoistic in our behaviour, the uniqueness of our conscious thinking, is actually a brain function developed by the unconscious to help us survive as a social group not as an individual. The 'WE' created the 'I'! How contradictory this sounds in the light of the profound individualism we observe in our organizations and personal lives. According to Halligan and Oakley (2015), our unconscious mind broadcasts all information and decisions to our conscious mind that then creates a personalized construction necessary for developing adaptive strategies in the real world – such as predicting the behaviour of others, disseminating selected information and being able to adjust perceptions based on external stimuli. And they are not alone in believing so. Neuroscientist Michael Graziano has suggested the Attention Schema Theory to argue that we have our consciousness in order to detect the consciousness of other people and thus to be able to make assumptions about their behaviour. He stated in a recent interview (Guzman, 2015):

> We perceive awareness in other people. And this is crucial to us as social animals. You know, it's not enough to look at another person and think of that person as some kind of robot where you have to predict what the next move of the robot might be. We have this kind of intuitive or gut impression that the other person is aware, has a mind, an inner experience. And we use that to help us understand other people, to help predict their behavior, to interact with them better. And we do this constantly. In fact, we're very much wired to see awareness in other things.

We are wired to use our attention to unconsciously detect signs in the other person's verbal and body language in order to better understand their state of mind and act accordingly. This is why it is called the Attention Schema Theory: we use part of our attention span to pick up patterns in other people's thinking, feeling and doing. Another Professor of Psychology and author of the book *The Neuroscience of Human Relationships: Attachment and the development of the social brain* (2006), Louis Cozolino, emphasized that 'the brain and the body are biological organisms. They are social organisms, such as the neuron in the brain is a social organism. It needs to connect with other neurons' (Sullivan, 2015). If people, and neurons, fail to create adequate connections across the social domain or across the brain respectively, they can be isolated or even rejected from the social and neuro system respectively with probably negative outcomes. As neurons need to create new pathways for the brain to remain healthy and to thrive, people need to create new relationships in order to remain healthy and to thrive. In order for neuroplasticity (as described earlier in the book) to work and to create novel neural pathways, our brain needs to be adequately stimulated. If this does not happen we have less chances for growth. Cozolino concludes, quite rightfully we may add, that: 'We're only beginning to realize that we're not separate beings, that we're really all just members of a hive, and it may take centuries to make us realize that we're much more interconnected than what we realize now' (Sullivan, 2015).

Isn't that true for professionals and organizations as well? The more a manager, a company, an industry, even a whole country gets disconnected from its overall environment, the more all these entities become ineffective and eventually decline. Our brain is primarily a connecting, interacting, trusting and cooperating organ and there is the evolution of human kind to provide the testimonial for it. Actually, our species' unique ability to form multi-layered social relations and to collaborate within highly complex and coordinated group activities with genetically unrelated individuals makes the single most important difference for us. Curtis Marean, Director of the Institute of Human Origins at the Arizona State University, believes firmly that Homo Sapiens' extraordinary ability to cooperate, what he calls hyperprosociality, is not a learned tendency but a genetically encoded trait and that this is what helped our species dominate against other related species, such as the Neanderthals (Marean, 2015). Although cooperation is also observed in primate species, our unique ability to collaborate in large, well-organized groups, by employing a complex morality competence based on reputation and punishment, was what gave the edge to humankind (De Waal, 2014).

It might come as a surprise but what really separates us from other species is not so much our superior reasoning but our social skills. Studies (Tomasello, 2014; Van der Goot *et al*, 2014) have shown that chimps and young human children perform equally on typical IQ tests but human offspring do much better on tests related to social-cognitive skills, like learning from each other (Stix, 2014). So, when someone brags about the power of competition in fuelling personal, economic and social growth, just remind them that without our brain's propensity for cooperation, mutual learning and social fairness we would still be in caves, even totally extinct! As Martin Nowak, Professor of Biology and Mathematics at Harvard University and Director of the Program for Evolutionary Dynamics, put it, 'people tend to think of evolution as a strictly dog-eat-dog struggle for survival. In fact, cooperation has been a driving force in evolution' (Nowak, 2012).

Leaders in modern organizations cannot go on with a mental state of fierce competition or a feeling of isolation from their social surrounding. On the contrary, wholeheartedly embracing collaboration and cooperation, alongside the necessary competition and unavoidable conflicts, will advance them to true brain adaptive leaders.

I know what you think

The ability of our brain to understand other people's mental states and to use this information to predict their behaviour is really astonishing. It is also an extremely helpful tool for managers and leaders in all organizations around the world. Neuroscience and psychology call this ability Theory of Mind (ToM) and it represents a core skill of our social brain. The foundations for this theory were laid down in the late 1970s when David Premack and Guy Woodruff asked in their seminal paper 'Does the chimpanzee have a theory of mind?' (Premack and Woodruff, 1978). They explained that:

> An individual has a theory of mind if he imputes mental states to himself and others. A system of inferences of this kind is properly viewed as a theory because such states are not directly observable, and the system can be used to make predictions about the behavior of others. As to the mental states the chimpanzee may infer, consider those inferred by our own species, for example, purpose or intention, as well as knowledge, belief, thinking, doubt, guessing, pretending, liking, and so forth.

Simply put, ToM is our species' advanced capacity to understand that we have cognitive, emotive and behavioural processes in our heads and that those are separate from the similar processes in other people's heads. In essence, it is a 'theory' because we can only make inferences of what happens in ours and other people's minds since we do not see directly into brains. So, we personally develop 'theories' or 'possible explanations' of what other people think, feel and are about to do. As we explained earlier, the domination of our species versus others depended on our brains' ability to be conscious about our own and other people's actions. Consequently, we can say the ToM is the main purpose of having a consciousness.

ToM is a complex system. It combines different brain areas responsible for memory, attention, language, executive functions, emotion processing, empathy and imitation, and it is profoundly dependent on the multi-layered interactions between brain development and social environment (Korkmaz, 2011). Interaction between brain and human environment is crucial. Isolation is fatal. Some children with cognitive and developmental disorders, such as autism and schizophrenia, have difficulties with ToM (Korkmaz, 2011). Understandably, ToM engages many brain regions in our cortex such as the prefrontal lobe related to the perception of emotions and the temporal lobe related to recognition of faces, and in deeper structures, such as the amygdala (see, for example, Martin and Weisberg, 2003). The fact that the amygdala is involved indicates the guarding role of ToM in our well-being. If the amygdala is hyperactive when engaging in ToM then it has probably detected danger in other people's potential actions and an avoid reaction is triggered. If not, an approach reaction is neurologically allowed and we can proceed to cooperation (you can find more on the amygdala and trust later in this chapter).

The healthy function of ToM is usually tested by specific tasks that reveal whether children, or adults (in a greater degree), can recognize correctly that other people can hold a false belief about something. That is, a belief that diverges from reality and can lead to a wrong action. The most famous of such tests involved two children, Sally and Anne, and was first proposed in the early 1980s by Wimmer and Perner (1983). Imagine two children, Sally and Anne, sitting side by side and holding one box each. Sally also holds a stone, which she places into her box and then she leaves the room without her box. While Sally is away, Anne opens Sally's box, takes the stone out and places it in her own box. Upon Sally's return, participating children (of various ages each time) are asked to predict in which box Sally will look for her stone. If the answer is in her own box then the children correctly recognized that Sally has a false belief. If they answer that Sally will look into Anne's box then they failed the test as they did not recognize Sally's individuality of mind, a mind that is separate from their own minds, holding opinions separate from reality. Sally does not know that Anne put the stone in her box, as the children themselves know, and thus mistakenly looks into her own box for the stone.

This and other false belief tests in ToM research showcase the importance of mind reading for human understanding, learning and collaborating – mind reading, not in a supernatural way, but in a very human, brain-based ability to hypothesize what another person is thinking. Shared intentionality, as the famed psychologist Michael Tomasello calls it, is necessary in developing a common ground between members of a group, helping them to achieve joint attention and ultimately to act cooperatively towards a collective goal (Tomasello and Carpenter, 2007). Mind reading, or intention understanding as it is also known, is highly developed in healthy humans and played a vital role in our evolution and dominance as a species.

The implications of ToM for the leadership brain are profound. Leaders play a central role in planning, in overseeing tactical implementation, in conversing, debating and negotiating both internally and externally with a large number of people, and in motivating and directing colleagues and employees. Their ability to understand the state of mind of the people across the table (or even of those absent) and to decode their intentions is key to successful decision making. Otherwise, leaders are deciding and acting in the dark. But what can we do to improve our mind-reading skills? How can we make sure this 'sixth sense' works best for us?

Let's start by abolishing the assumption that adults are by default perfect in mind reading compared to kids and that this ability comes effortlessly whenever needed. Yes, we do score better than kids at tests designed to reveal egocentricity vs perspective taking but not as dramatically better as expected. A relevant study found that children failed to take the other person's perspective in 80 per cent of the requested tasks but adults in 45 per cent of them (Dumontheil *et al*, 2010). This is not a spectacular difference. Matthew D Lieberman, Professor in the Departments of Psychology, Psychiatry and Biobehavioral Sciences at the University of California, Los Angeles, commented on this study, saying:

> Yes, adults have the capacity to mentalize well, but as this study shows, they don't apply this tendency reliably. This is probably because the brain regions that support accurately mentalizing require effort to work well, and we are wired to be mental couch potatoes whenever we can get away with it. We mentalize a great deal, but this doesn't mean we always do it well or that we can't learn to do it better. (Lieberman, 2013)

Mentalize is the ability to understand the behaviour of others as a product of their mental state (Frith and Frith, 1999). According to Lieberman, the problem is that too often we allow our mental shortcuts to jump to conclusions about other people's preferences and intention by using as a main input what we ourselves prefer and intend to do. We base our mind-reading on our own mental state too much, making serious mistakes in our judgements just because we do not take the time to really consider the situation. He has also found that spending brain power on non-social mental tasks prior to engaging in ToM lowers our ability to be accurate. On the contrary, the more 'socially' we think before a ToM task, even the more we put our brains in a default mode (relaxing or daydreaming modes), the better we become at it (Lieberman, 2013).

We can personally, and we are sure you can too, find numerous examples from everyday working life when people, including us, fall into those ToM-impairing traps. How many times do we think that we empathize with others or that we put ourselves 'in someone else's shoes' when in fact we are just projecting our own mental state onto them? Actually, a recent organizational study found exactly that. In a series of experiments with 480 experienced marketing managers, Johannes D Hattula and his associates found that those managers who try to empathize with their customers before predicting what those customers wanted, performed very poorly in getting it right (Hattula *et al*, 2015). In their attempt to empathize with customers, managers in fact accelerated self-reference because taking the customer perspective activated their own private consumer identity. Their own personal consumption preferences were reflected in the predicted ones. Furthermore, their self-referential preference predictions made them less likely to use market research results, highlighting the fact that wrong mentalization gets even worse with the false confidence that we are good at it. Last but not least, they found that the managers instructed explicitly not to empathize with customers performed better.

Nicholas Epley, Professor of Behavioral Science at the University of Chicago's Booth School of Business, reinforces in his book *Mindwise* the point of not trying to empathize with others by lazily fantasizing about their needs. Instead, he suggests revealing other people's actual perspectives by direct interaction with them, or what he calls 'perspective getting' (Epley, 2014). He uses many examples where organizations or individuals understood better what other people wanted by interacting with them and by directly probing their intentions. Such an approach produces much better results than just staying inside an office and assuming other people's needs based on past knowledge, personal experiences and our own preferences. However, he did admit that getting other people's perspectives directly and using a verbal communication approach, can be problematic. So, he recommends that when interacting directly with the people you want to mind read, first try to build rapport so they feel comfortable, second, be less speculative and elusive in the discussion as possible, and third, be a good listener and paraphrase often to make sure that you understand correctly what the other person is saying.

Finally, engaging in ToM in groups can be more successful than trying it alone. This means that if you and your colleagues try to mind read together other people (such as customers, suppliers, people from other departments etc) you will probably do it better than if you tried it by yourself. In a recent study, researchers showed video statements to participants asking them to spot individually and in groups, truth or lies. Results indicate that while individuals and groups were equal in spotting truths, groups were consistently better than individuals in spotting lies – both white, innocent lies and high-stake, intentional ones. According to the researchers:

> This group advantage does not come through the statistical aggregation of individual opinions (a 'wisdom-of-crowds' effect), but instead through the process of group discussion. Groups were not simply maximizing the small amounts of accuracy contained among individual members but were instead creating a unique type of accuracy altogether. (Klein and Epley, 2015)

This is the true expression of healthy brain synergy within a team. The whole is greater than the sum and a few minds interacting together produce better outcomes than individual and isolated thinking. Isn't this the quintessential goal of teamwork?

The main lessons for the brain adaptive leader in mind reading are clear:

- Lazy assumptions of what other people think, feel and will probably do are simply projecting our own state of mind and are doomed to fail.
- Actively trying not to project our own situation upon others helps a lot in taking the right perspective.
- Doing our homework can never be replaced just by mind reading. Market, organizational, industrial and personal research should always be utilized.
- Trying not to engage in tiring mechanical, technical, analytical thinking before engaging in ToM will improve the results of mind reading.
- Relaxing, meditating and trying to think about people can improve ToM results.
- Engage directly with others to get their perspective (rather than just guessing it) by creating a comfortable situation, being as clear and transparent as possible, listening actively and making sure you understood the other person correctly.
- Practise ToM collectively with your team. It can be more successful than doing it alone, especially in spotting deception which, if unnoticed, can cost the organization a lot.

The mirrors in our brains

The concept in the area of mind reading, empathy, imitation and social bonding that has stirred global heated debates in neuroscience and psychology over the last 20 years is the concept of the so-called mirror neurons. Mirror neurons are the neurons that activated when a human acts as well as when a human observes the same action performed by another human (Keysers, 2010; Rizzolatti and Craighero, 2004).

In 1992, Italian researchers published a study showing that a macaque's neurons fired, not only when the monkey was performing an action but also when this monkey was observing another one doing so (di Pellegrino *et al*, 1992). Since then, numerous studies, opinions and forums appeared either supporting and expanding on the mirror neurons theory or doubting what mirror neurons actually do, but not questioning the existence of mirror neurons (Marshall, 2014). Celebrated supporters of the wider role that mirror neurons play in our individual lives, group workings and even in our cultures include neuroscientists VS Ramachandran (2011), who believes that the discovery of mirror neurons is equal in significance to the discovery of DNA, and Marco Iacoboni (2009) who claims that mirror neurons are the special brain cells that can finally help us answer centuries-old philosophical and scientific questions about how we successfully organize societies and cultures. Opponents are also quite vocal, with Hickok (2009) identifying a number of empirical problems associated with mirror neurons research, and Kilner (2011) highlighting that the main function attributed to the mirror neurons, action understanding, has alternative neural pathways to materialize itself. Mirror neurons are here to stay and leaders should be aware of their existence and suggested impact regardless of whether this concerns motor skills-related functions or deeper emotional ones such as empathy and social bonding

In the leadership literature arena, Daniel Goleman and Richard Boyatzis published an influential article in the *Harvard Business Review* titled 'Social intelligence and the biology of leadership' in 2008, which included a whole section on mirror neurons. For them, mirror neurons play an important role in leadership effectiveness since followers of the leader will copy or mirror the leader's mental state based on how the leader expresses it. They mention a study where employees presented with negative performance feedback, accompanied by positive emotional signals, reported feeling overall better than employees presented with positive feedback but through negative emotional signals. So, leaders should be very careful what emotions and mood they show, mainly through their body language, because this profile will be mirrored unconsciously by the people around them (Goleman and Boyatzis, 2008).

We would add in the same direction, based on what is discussed in this chapter, that there is a high probability that people will also mirror a strong leader's posture, gestures and overall body language as well as the leader's overall mindset. Furthermore, if you find the mindset of your own superiors or/and colleagues incompatible with the purpose of the business then try not to mirror them yourself. Imitation works best when what is imitated is positive to the individual and to the team.

Imitation is a powerful brain mechanism that contributes directly to our development, learning, group assimilation and cultural homogeneity, cooperation and even to innovation, at various stages in our lives. It is a fundamental and hard-wired function of our brains and babies as young as 45 minutes old have the capacity to imitate facial expressions (McGilchrist, 2012). It seems that imitation, or mimesis as it is scientifically known:

> ... let us escape from the confines of our own experience and enter directly into the experience of another being: this is the way in which, through human consciousness we bridge the gap, share in what another feels and does, in what it is like to be that person... It is founded on empathy and grounded in the body. In fact, imitation is a marker for empathy: more empathic people mimic the facial expressions of those they are with more than others. (McGilchrist, 2012)

Leaders and managers should not behave like imitation does not exist, thus ignoring their responsibility in shaping other people's moods, attitudes and behaviour. Lead-by-example is now more relevant than ever due to the amazing attribute of some of our neurons to copy/paste automatically and naturally what happens in front of our eyes. The power of imitation shapes the culture of our organization, departments and teams in ways that go unnoticed to the non brain-adapted managerial mind. We firmly believe, based on our worldwide experience, that mimicking is really the easiest way of shaping a company's or unit's working atmosphere as well as for assimilating desired values in groups. We just need to 'walk the talk'; always and in a consistent manner. Mirror neurons and mimesis can work miracles if we choose to acknowledge them and use them in the right direction daily in our working lives!

Human connectivity

Leaders are not, or should not be, islands. They should be actively connected to people inside and outside their organization, with those close to them but also with those further away from their immediate personal circle. We see the creation, maintenance and constant development of powerful human networks as an indispensable aspect of modern leadership in all organizations. This is because our brain, as a primarily information-processing organ (Sullivan, 2015) needs the right sources and input of information in order for it to create great output in the form of ideas, emotions and actions. Simply put, the quality and quantity of brains your brain interacts with will determine, to a high degree, the quality and quantity of your own brain performance. As a true social animal, the brain adaptive leader understands the importance of human connections and networks and works tirelessly to build bridges that matter. But which connections are the ones that matter the most in the context of brain-calibrated leadership and the complex business and organizational environments we operate in?

The classic social network theory of 'the strength of weak ties' developed by American sociologist Mark Granovetter in the beginning of the 1970s and advocated strongly since then, is an excellent mental tool for understanding and making better use of our personal connections to other people. Granovetter tried to provide a comprehensive explanation of human networks, applying simultaneously micro and macro elements of sociological analysis, and suggested that we need to urgently look at the ties between people that are not strong, in the sense of relationship proximity (that is, close family members, friends and colleagues). In other words, we need to turn our attention, at least in professional terms, to weak ties, meaning ties with those with whom we have distant or/and indirect connections (that is, acquaintances, distant friends, colleagues and distant family members). The latter have been largely ignored by sociology until 1973. The argument goes that although strong ties are invaluable in providing everyday support, it is the weak ties that inject necessary new ideas, new experiences and new practices in a group of people. Thus, this group must establish links with people who have different characteristics; that is, live and operate in distinct socioeconomic levels. This applies equally to individuals, towns and cities, companies and organizations and even to whole industrial districts with thousands of SMEs collaborating and competing with each other in order to survive in global competitive markets (Dimitriadis, 2008). Fresh ideas, strategic information, out-of-the-box innovations, change initiatives, alternative perspectives and disruptive technologies are rarely the product of tired, over-squeezed and over-familiar relations within groups of people working together too closely for too long. They are born somewhere else in the socio-economic sphere of human activity and need to be noticed, considered and ultimately embraced by a group of people in order to spawn progress. Think of your organization as a collective brain: if isolated from new stimuli and from the challenges posed by new information it eventually degenerates and becomes dysfunctional. On the contrary, if it creates and uses appropriately 'weak' links with the outside world in multiple directions then it is kept healthy and neuroplasticity pushes constantly for new pathways, opportunities and capabilities.

In order for weak ties to work best they need to be qualitatively different from a simple 'friend-of-a-friend' connection. If this new link is actively used for producing new information for the individual and the group then it is not just a distant connection but a very helpful one; it is a bridge (Granovetter, 1983). Based on Granovetter's ground-breaking for its time approach, and on our experience on the topic, we have developed a straightforward matrix to help us explain to leaders and managers the importance of building and maintaining the right bridges. We call it the leadership network connections matrix (see Table 7.1).

TABLE 7.1 The leadership network connections matrix

Type of connection	Power of connection	
	Dynamic/productive	*Static/non-productive*
Strong/close	Active operational support	Potential dead-end
Weak/distant	Active strategic bridge	Potential beneficial link

Weak ties that are active and productive, supply crucial strategic information to the leader and to the team, and thus are instrumental in introducing change, innovations and evolution. Leaders should be proactive in creating such connections since this is a core responsibility of the leadership role in any organizational level. For example, they need to attend events such as conferences, conventions and fairs with an open mind and a proactive approach to be able to create new contacts. That's the first step. The second step is to invest personal time to nurture those contacts in order to keep them active and be able acquire from them useful information. Weak ties that exist but are not utilized are potential bridges and although not much time should be spent on them, they should remain within the leader's mental radar and reach. Concerning strong ties, if they are positively contributing to the leaders' and the team's operational life then they should be maintained and nurtured because they form a crucial everyday support network.

If they are passive and negative though, the leader should adopt ways to challenge them and make them reconsider their role within the group and/or team. One of the methods that we usually suggest in our consultancy projects is called small and controllable earthquakes. Leaders need to challenge passive and negative people in order to make them reconsider their attitudes by applying various methods that are critical enough to attract attention, but small in order to be under control. This sometimes requires leaders' personal involvement to guarantee the success of the method. But this is a key role of leaders anyway, to be personally involved. Isn't it?

A brain adaptive leader cannot afford to be trapped in counter-productive close relations and without distant (socio-economically speaking), but powerful 'weak' bridges. All those micro and macro relations form our true human network in which we live and work. Their impact on our success or failure and on our emotional and mental states is now revealed to be profound. So profound that even researchers studying human networks for decades were caught by surprise when confronted with the colossal impact of those networks on various human conditions.

Nicholas Christakis and James Fowler, in their highly appraised book *Connected: The amazing power of social networks and how they shape our lives*, took the world by storm, revealing the unexpected ways in which human networks influence who we are. Their main conclusions (Christakis and Fowler, 2010), summarized in three groups here, are discussed through the lenses of the BAL framework:

1 *We make networks*. We constantly, consciously or unconsciously, define our participation and position in human networks. To how many people are we going to be connected, strongly or weakly, and what role are we going to play in these relationships? Such actions, either as deliberate decisions or intuitive behaviours, will determine the experience that we will have in the network. We do not always choose the people we work with but we choose the people we are really connected with. Thus, leaders should never hide. They should purposely acquire central and active positions in multi-layered networks inside and outside their organizations, connecting to people in accordance with the Leadership Network Connection Matrix discussed above. Leaders make networks; they do not simply participate in them.

2 *Networks make us*. The number and quality of connections determines our own state of mind, well-being, habit formation and happiness levels. And this goes beyond the people we know. It also involves the

people that are connected with the people we know and that we have never met. Transitivity, the process of characteristics passing through different degrees of separation from people to people (through imitation) works for up to three degrees of separation. This means that the friend of a friend of a friend of yours, that you have never met, influences your happiness and your health habits more than you ever thought was possible. In other words, a decision that a person will make with whom you are indirectly linked (practically meaning, you do not know him/her personally) can influence the decision of a person who you know that in turn can influence a decision of yours. Once again chaos theory in action, since this works best in complex and rich networks and not in connections arranged as a simple line. Social reinforcement from multiple connections is important for norm and behavioural change. Thus leaders need to be cautious about the portion of their behaviour attributed to network transitivity. Thus, observe and monitor behavioural patterns in your team and explore their origin. If positive, reinforce them. If negative, try to disassociate through increased awareness, meaning that noticing them will most probably make you alter you behaviour accordingly. Not noticing them will allow for the situation to persist. On the other hand, to create desired norms a leader should take advantage of transitivity and increase interactions that promote those norms. This is possible by introducing more people who behave accordingly in different levels of the organization and even from the outside of the company. Leaders are aware of network influences and use them to their advantage; they do not just blindly go along with them.

3 *Networks are alive (and often independent of us)*. Complex, direct and indirect relations in dense human networks have collective characteristics that are greater than the sum of their parts. Networks exist and change independently from each individual contribution and remain active before and after each individual participation. However, they also have emergent properties that are attributes deriving from the multilayered interactions and interconnections of its members. In that way, cultures depend less on individual actors and more on the interplay of many actors that will consciously or unconsciously interact with its norms and with each other. Leaders of larger organizations are in the position to influence their culture in a slower manner and by using more tactics than those in a smaller one. In both cases though, leaders should be aware of the fact that cultures

are complex living entities that do not change by simply changing a rule or a person. They move in unison. In this respect, change agents, opinion leaders and influencers in general cannot achieve great results if previously people did not have the time to adjust in the new situation. In other words, a change, especially a radical one, cannot be applied on the spot (at its emergence), but after a period of time when people become more familiar and aware of it. This is why dramatic changes in opinions, attitudes and behaviours within a culture rarely happen without first making sure that people collectively have reached a critical point of accepting those changes (Watts and Dodds, 2007). Leaders orchestrate changes; they do not necessarily bring them about alone.

In a hyper-connected environment created by social media and the online world, Christakis and Fowler (2010) believed that social networking changes in the digital domain but also that its fundamentals remain the same. It is true that Facebook, LinkedIn, Twitter, Instagram and other social media allow us to make more connections faster and easier than ever before. We can never use avatars to hide our true identity. But are all those connections active support networks and real bridges? We can't say for sure because it depends on the connection. Social media have accelerated connections and often provide real value in transitivity. But they do not automatically mean active support or bridges. As Dr Julian Ney, a social intelligence expert, convincingly argued in her 2014 TEDx speech 'Connectedness and the digital self' we need to put connectedness, a deeper psychological link, back into the discussion of online connections (Ney, 2014). We, too, believe that being connected to people or organizations through social media does not necessarily mean engaging mentally and emotionally with them and thus cannot be considered a priori as active human network connections. Undoubtedly, leaders and managers need to be present in social media (and we will discuss in the next chapter how this should be done) and create potentially active links, but they still need to nurture and deliberately manage connections to people online and offline in order to experience positive and dynamic relations with them.

Ultimately, we see ourselves, as Christakis and Fowler (2010) do, as members of superorganisms: the living, breathing human networks we belong to. As such we ought to contribute decisively to them, to shape them but also to be aware of how they shape us. Our brain leadership potential depends on that.

The art of human connectivity for leaders

How can modern leaders become better in connecting with other people inside and outside their organizations? Ori and Rom Brafman suggest that in order to accelerate our ability of creating instant deep connectivity, something they call 'clicking' with other people, we need to increase our vulnerability (Brafman and Brafman, 2010). Although this may sound counterintuitive, we can connect instantly to the deeper emotional and mental states of people if we allow ourselves to look open and vulnerable. In that sense, vulnerability induces instant trust and instincts of care in other people. Of course, this has to be expressed appropriately because over-vulnerability can give unwanted weakness signs. Brafman and Brafman (2010) refer to a vulnerability continuum present in human interaction. On one end we have non-vulnerability and at the other high vulnerability. The five distinct stages on this continuum, as revealed though everyday conversations, are the following:

- *Phatic stage*. We speak with social niceties, words void of any real emotional charge such as 'how are you?' and 'nice to see you'. The purpose here is not to extract any specific response but to decrease any possible problems in our interaction.

- *Factual stage*. We speak with actual facts and data that do not elicit any subjective opinions, such as 'I am an engineer' or 'I live in Europe'. Such straightforward observations reveal a bit more about us but not in a deep emotional way.

- *Evaluative stage*. We speak with personal views or opinions about people, places and situations, such as 'I like this plan' and 'this product is amazing'. Here we take a limited social risk because the other people might not agree. Still nothing dramatic though.

- *Gut-level stage*. We speak in emotionally revealing statements like 'I am so sad Mark is leaving the company' or 'I am really glad I hired you'. Those statements are more risky and leave us open to criticism and to emotional disagreement. However, they are the kind of statements made to people closer to us, people in our circle of trust.

- *Peak stage*. This is the most revealing stage where we open our heart and mind, and speak about our inner feelings, deep fears and wildest hopes. Statements such as 'I really wonder how we are going to make it this month with so low numbers. I fear for my job although it is unconceivable to me that I am in danger having spent my whole life in this office. Do they not care about the past; am I not important to them beyond numbers? This is scary...'.

The first three stages are more transactional while the last two are more connective. It is our duty to assess a situation and decide which approach we have to employ in order to connect faster and deeper with the people we talk to. Starting at stage 5 with someone we just met is not advised. But moving gradually from the first three to the last two ones, even during a single conversation, can help us show some degree of vulnerability and connect instantly with people.

The way by which leaders can mould their behaviour in order to become more connectable to others in the long run, based on neuroscientific and psychological findings, is described in the best-selling book *The Charisma Myth* by Olivia Fox Cabane. Cabane (2013) summarized the characteristics of charismatic leaders into three clusters. For us, these clusters correspond perfectly to the three brains approach presented earlier in this book.

- *Cluster A: Presence.* Portraying the feeling that you are 100 per cent focused on the issue and the person (or persons) in front of you. We have all experienced situations in our jobs where we felt that the person in front of us was not actually following or listening to what we were saying. Not being focused makes other people feel distant and even betrayed. Great leaders make people feel noticed and important regardless of their position. That's absolutely necessary for creating meaningful connections inside and outside the company. This cluster refers to the older, more primitive brain because it is through our body language, posture, gaze and tone of voice that people will understand, consciously or unconsciously, that we are there for them. The automatic brain will pick up signals portraying presence or not even when you talk to people over the phone.

- *Cluster B: Warmth.* Empathizing with people and showing that you really care about their situation is the second cluster. Goodwill, compassion and acceptance are core processes here, since through warmth we make people feel comfortable, looked-after and protected. This refers to the emotional brain, and especially the mid-brain systems of approach and avoidance, as discussed in previous chapters, since our own stance will make people feel safe and get closer to us or feel threatened and get away from us. Great leaders give strong and clear signs of compassion and thus create meaningful connections that last.

- *Cluster C: Strength.* This is related to perceptions of power and authority and, although it does include some elements of the primitive and mid brains, it is mostly rooted in the neocortex: it refers to our ability to do something specific for the people we want to connect with. Think about it. Being present is great and being compassionate is even greater. But real leadership requires you to be able to take specific decisions and actions that will practically improve your team's performance. Your ability and willingness to demonstrate in action that you can help is a matter of advanced analytical and technical skills. Nobody can connect to a leader who is unwilling or incapable of demonstrating (proving) her or his expertise when most needed.

New voices are added on the importance of warmth and strength both in our personal lives (Neffinger and Kohut, 2014) and in brand management (Malone and Fiske, 2013). Executive presence has also become an area with its own advocates and expanded scope (see for example Hewlett, 2014). Although we can rarely be perfect in all three, we believe that actively noticing our performance in all of them will help us improve the way we connect and maintain relations.

Trust the amygdala and your brain chemicals

Trust and the amygdala do not usually go hand in hand. This is because whenever there is heightened activity in the amygdala we trust less and when the amygdala is quiet we trust more. We do though need to trust, in the sense of listening to, the amygdala on this.

The amygdala is the usual culprit to accuse when people overact with fear and anger in a situation, and thus fail to manage their emotions effectively and deeply connect with others (see Chapters 3 and 4). However, the amygdala plays a crucial role in regulating the level of trustworthiness of other people. Simply put, without the amygdala we would trust everyone without any limits and with catastrophic results for us and our organizations

(Fox, 2013). The amygdala, together with the insula and other brain centres, is a key neural area activating our alarm system when confronted with untrustworthy-looking faces (Winston *et al*, 2002). This hard-wired cautionary system, provided by the amygdala to protect us in risky situations, is absent if the amygdala is dysfunctional. Actually, people with deactivated amygdala have their natural guard down and are ready to trust anyone with everything (Adolphs *et al*, 2002). It is the amygdala that fires up in order to detect fear and anger in other people and even to differentiate between those two emotions (Fox, 2013). Since faces expressing those two emotions are usually deemed as less trustworthy, the absence of a healthy amygdala stops us short in taking precautionary measures when interacting with them, making people without amygdala even ready to give their credit card details out to random strangers if asked to do so (Fox, 2013). This is happening since the amygdala has an important role in evaluating the emotional valence of stimuli. For example, people with amygdala injuries have difficulty in learning associations between environmental stimuli and emotional states. Thus, they may fail to learn that a stimulus predicts reward or danger, they may also fall in social rank, or show decreased affiliative behaviour (Blair, 2001). Thus, we should be extremely grateful for a fully functioning amygdala in separating out which people to trust and not to trust. If you experience feelings of avoidance when meeting people, explore those feelings before accepting or rejecting them. It might be your healthy amygdala sending you justified signs of caution or it could be an oversensitivity of your avoidance system making you less approachable. Learning is a core skill of brain adaptive leaders.

In the opening case to this chapter, our protagonist had nurtured a hyperactive amygdala that did not allow her to connect meaningfully to the most important people around her at work. In our modern, dynamic and complex global environment, more and more employees choose to work for companies they feel personally and emotionally connected to, as we saw from the 2015 results of *Fortune*'s Top 100 Best Companies to Work For in Chapter 5.

Apart from the amygdala, there are two important brain chemicals involved in developing trust between ourselves and people in our human network. One is oxytocin and the second is vassopressin. The latest studies suggest that oxytocin plays a central role in organizing our social behaviour because it increases our propensity to trust others, approach and connect with them, and create meaningful relationships (Levitin, 2015). We can significantly increase the flow of oxytocin by playing and enjoying music together with other people (Levitin, 2015) or even by simply giving them a big hug eight times per day (Zak, 2011). The second neurochemical is called vasopressin and is crucial for regulating bonding, sociability and stress responses between humans and other mammals (Levitin, 2015). This chemical has been found to contribute extensively to retaining relationships by decreasing the defect option (Young, 2003). It increases close bonding between people by attaching excitement to the relationship but it can fade out over time. Thus the challenge is to identify ways to reinvigorate the relationship when we detect emotional detachment signs in others (Young, 2003). The leaders' role is to keep their connections alive, dynamic and productive and this means infusing excitement and positive surprises in their relationships to be able to maintain workable levels of neurochemicals facilitating bonding.

The exclusion of humans at work, in terms of machines replacing humans in many typical job positions, has been prophesied by many but most notably by Jeremy Rifkin in his classic *The End of Work* book back in the mid-nineties (Rifkin, 1995). More recently, the *Harvard Business Review* ran in its June 2015 issue a front page with the picture of a robot and with the title 'Meet your new employee: how to manage the man-machine collaboration'. Although machines play an increasing role in many areas of business and organizational activity, the central stage is still set for humans. In his 'Humans are underrated' article in *Fortune Magazine* adapted from his same-titled book, Geoff Colvin explains that humans will remain in charge for the foreseeable future and that only humans can satisfy deep interpersonal needs so profoundly important for engagement inside and outside our organizations (Colvin, 2015). The fact that empathy levels, as measured by 72 studies of 14,000 students in the United States, have fallen by more than 10 per cent in the last 30 years, shows that although empathy is in great demand ('empathy is the critical 21st-century skill' for Meg Bear, the Group VP for Oracle) its supply is shrinking and this asks for a dramatically new turn in how we prepare workers, managers and leaders today (Colvin, 2015). Colvin concludes, and we couldn't agree more, that:

> For the past 10 generations in the developed world, and shorter but still substantial periods in many emerging markets, most people have succeeded by learning to do machine work better than machines could do it. Now that era is ending. Machines are increasingly doing such work better than we ever could... Fear not... you'll find that what you need next has been there all along. It has been there forever. In the deepest possible sense, you've already got what it takes. Make of it what you will.

> **Boost your brain**
>
> Ask your team to write down all the necessary skills needed between you all to succeed in your goals. Rate them from 1 to 5 based on how effectively they can be done by machines (closer to 1) or by humans (closer to 5). Calculate the total scores and averages for all skills and use the results to initiate a group discussion about the unique characteristics of human contribution. You can also discuss how machines can assist but the main purpose is to identify what make humans capable in creating an engaging environment and in developing social connections that matter. How can the team improve? Use the materials in this (and in previous chapters) to progress the discussion and to build consensus.

Keep in mind

Leaders are not islands. On the contrary, they take central positions in organizational and business networks in order to facilitate relations that will help them achieve their goals. However, they need to recalibrate their brains to better understand the social behaviours of themselves and others, and to constantly connect deeper with people both inside and outside their organizations. Fully appreciating the socially based origins of consciousness, adopting a 'collaborate first and then reciprocate' attitude in business interactions, utilizing imitation as a cultural creation and maintenance tool, establishing and managing extensive human networks, managing the art of clicking and of demonstrating presence, warmth and strength, and making sure that the amygdala and a few brain chemicals are in a careful collaborative mode, are necessary conditions towards adapting our brain for networked leadership. The era of the machines also paradoxically brings the era of human beings. Be the first to thrive by cultivating in your brain what makes us uniquely human!

References

Adolphs, R, Baron-Cohen, S and Tranel, D (2002) Impaired recognition of social emotions following amygdala damage, *Journal of Cognitive Neuroscience*, **14** (8), pp 1264–1274

Axelrod, R (1984) *The Evolution of Cooperation*, Basic Books, New York

Brafman, O and Brafman, R (2010) *Click: The forces behind how we fully engage with people, work and everything we do*, Crown Business, New York

Blair, HT (2001) Synaptic plasticity in the lateral amygdala: A cellular hypothesis of fear conditioning, *Learning and Memory*, 8 (5), pp 229–242

Cabane, OF (2013) *The Charisma Myth: Master the art of personal magnetism*, Portfolio Penguin, London

Capraro, V (2013) A model of human cooperation in social dilemmas, *PLoS One*, 8 (8), e72427

Christakis, NA and Fowler, JH (2010) *Connected: The surprising power of our social networks and how they shape our lives*, HarperPress, London

Colvin, G (2015) Humans are underrated, *Fortune*, European Edition, 1 August, 172 (2), pp 34–43

Dawkins, R (1976) *The Selfish Gene*, Oxford University Press, Oxford

De Waal, F (2014) One for all, *Scientific American*, Special Evolution Issue: How We Became Human, 311 (3), September, pp 52–55

di Pellegrino, G, Fadiga, L, Fogassi, L, Gallese, V and Rizzolatti, G (1992) Understanding motor events: A neurophysiological study, *Experimental Brain Research*, 91 (1), pp 176–180

Dimitriadis, N (2008) Information flow and global competitiveness of industrial districts: Lessons learned from Kastoria's fur district in Greece, in *Innovation Networks and Knowledge Clusters: Findings and insights from the US, EU and Japan*, eds EG Carayannis, D Assimakopoulos and M Kondo, pp 186–209, Palgrave Macmillan, London

Dimitriadis, N and Ketikidis, P (1999) Logistics and strategic enterprise networks: Cooperation as a source of competitive advantage, 4th Hellenic Logistics Conference, Sole – The International Society of Logistics – Athens Chapter (in Greek)

Dumontheil, I, Apperly, IA and Blakemore, SJ (2010) Online usage of theory of mind continues to develop in late adolescence, *Developmental Science*, 13 (2), pp 331–338

Epley, N (2014) *Mindwise: Why we misunderstand what others think, believe, feel, and want*, Allen Lane, London

Frith, CD and Frith, U (1999) Interacting minds – a biological basis, *Science*, 286 (5445), pp 1692–1695

Fox, E (2013) *Rainy Brain, Sunny Brain: How to retrain your brain to overcome pessimism and achieve a more positive outlook*, Arrow Books, London

Goleman, D and Boyatzis, R (2008) Social intelligence and the biology of leadership, *Harvard Business Review*, 86 (9), pp 74–81

Granovetter, M (1983) The strength of weak ties: A network theory revisited, *Sociological Theory*, 1 (1), pp 201–233

Granovetter, MS (1973) The strength of weak ties, *American Journal of Sociology*, 78 (6), pp 1360–1380

Guzman, IP (2015) What is consciousness? Dr. Michael Graziano and the Attention Schema Theory, *BrainWorld*, **6** (2), Winter, pp 46–49

Halligan, P and Oakley, D (2015) Consciousness isn't all about you, you know, *New Scientist*, **227** (3034), 15 August, pp 26–27

Hewlett, SA (2014) *Executive Presence: The missing link between merit and success*, HarperCollins, New York

Hickok, G (2009) Eight problems for the mirror neuron theory of action understanding in monkeys and humans, *Journal of Cognitive Neuroscience*, **21** (7), pp 1229–1243

Iacoboni, M (2009) *Mirroring People: The science of empathy and how we connect with others*, Picardo, New York

Keysers, C (2010) Mirror neurons, *Current Biology*, **19** (21), pp 971–973

Kilner, JM (2011) More than one pathway to action understanding, *Trends in Cognitive Sciences*, **15** (8), pp 352–357

Klein, N and Epley, N (2015) Group discussion improves lie detection, *Proceedings of the National Academy of Sciences*, **112** (24), pp 7460–7465

Korkmaz, B (2011) Theory of mind and neurodevelopmental disorders of childhood, *Pediatric Research*, **69**, pp 101R–108R

Lieberman, MD (2013) *Social: Why our brains are wired to connect*, Broadway Books, New York

Malone, C and Fiske, ST (2013) *The Human Brand: How we relate to people, products and companies*, Jossey-Bass, San Francisco

Marean, WM (2015) The most invasive species of all, *Scientific American*, **313** (2) August, pp 22–29

Marshall, J (2014) Mirror neurons, *Proceedings of the National Academy of Sciences*, **111** (18), p 6531

Martin, A and Weisberg, J (2003) Neural foundations for understanding social and mechanical concepts, *Cognitive Neuropsychology*, **20** (3–6), pp 575–587

McGilchrist, I (2012) *The Master and His Emissary: The divided brain and the making of the western world*, Yale University Press, New Haven

Neffinger, J and Kohut, M (2014) *Compelling People: The hidden qualities that make us influential*, Piaktus, London

Ney, J (2014) Connectedness and the digital self: Jillian Ney at TEDx University of Glasgow [video file], TEDx Talks, URL: www.youtube.com/watch?v=3QA8iy7sjT8, accessed 15 September 2015

Nowak, MA (2012) Why we help, *Scientific American*, **307** (1), July, pp 20–25

Poulsen, K (2014) No limit: Two Las Vegas gamblers found a king-size bug in video poker. It was the worst thing that could have happened to them, *WIRED*, November, pp 138–145

Premack, D and Woodruff, G (1978) Does the chimpanzee have a theory of mind?, *Behavioral and Brain Sciences*, **1** (4), pp 515–526

Ramachandran, VS (2011) *The Tell-Tale Brain: A neuroscientist's quest for what makes us human*, WW Norton, New York

Ridley, M (1997) *The Origins of Virtue: Human instincts and the evolution of cooperation*, Penguin, London

Rifkin, J (1995) *The End of Work: The decline of the global labor force and the dawn of the post-market era*, Putnam, New York

Rizzolatti, G and Craighero, L (2004) The mirror-neuron system, *Annual Review of Neuroscience*, **27** (1), pp 169–192

Stix, G (2014) The 'It' Factor, *Scientific American*, Special Evolution Issue: How We Became Human, **311** (3), September, pp 72–79

Sullivan, J (2015) Born to trust: The brain evolution as a social organism – a conversation with Louis Cozolino, PHD, *BrainWorld*, **6** (2), Winter, pp 50–53

Thaler, RH (2015) *Misbehaving: The making of behavioral economics*, Allen Lane, London

Tomasello, M and Carpenter, M (2007) Shared intentionality, *Developmental Science*, **10** (1), pp 121–125

Tomasello, MA (2014) *Natural History of Human Thinking*, Harvard University Press, Boston

Van der Goot, MH, Tomasello, M and Liszkowski, U (2014) Differences in the nonverbal requests of great apes and human infants, *Child Development*, **85** (2), pp 444–455

Watts, DJ and Dodds, PS (2007) Influentials, networks, and public opinion formation, *Journal of Consumer Research*, **34** (4), pp 441–458

Wimmer, H and Perner, J (1983) Beliefs about beliefs: Representation and constraining function of wrong beliefs in young children's understanding of deception, *Cognition*, **13** (1), pp 103–128

Winston, JS, Strange, BA, O'Doherty, J and Dolan, RJ (2002) Automatic and intentional brain responses during evaluation of trustworthiness of faces, *Nature Neuroscience*, **5** (3), pp 277–283

Young, LJ (2003) The neural basis of pair bonding in a monogamous species: A model for understanding the biological basis of human behavior, in *Offspring: Human fertility behavior in Biodemographic Perspective*, eds KW Wachter and RA Bulatao, pp 91–103, The National Academy Press, Washington DC

Zak, P (2011) Paul Zak: Trust, morality – and oxytocin? [video file], TEDGlobal Talks, URL: www.ted.com/talks/paul_zak_trust_morality_and_oxytocin?language=en, accessed 17 September 2015

Brain communication, better persuasion

08

Why don't they follow our values?

The new corporate values are finally ready. Revising and changing the company's vision and values was on the top of the new CEO's agenda. As soon as she arrived in the company, she gathered her team and swiftly asked them to adopt new values that were more motivational and inspiring than the old ones. She believed that by changing the existing product and company-oriented values to customer and society-centric ones, she would also change the corporate culture. This was one of her key mandates when hired by the board to transform the company from a traditional, slow-moving corporation to a modern, fast-thinking business. She was convinced that values were an excellent place to start!

The corporate communications department worked closely with human resources and even marketing to identify and develop values that reflected the vision of the new CEO. It took them months of internal and external research, analysis and thinking, and with the help of external partners, such as agencies and consultants, they produced a set of modern-sounding, outward-looking, motivational values, such as people come first, customer focus, autonomous work, emphasis on cooperation and synergy, respect for the work of others, etc. The CEO was very happy. She now had corporate values representative of the type of company she wanted to make out of the existing one. Being proud of the new values, she could not stop talking about them in social meetings with peers from other companies. Many congratulated her and this reinforced the feeling of achievement and pride she had for the project.

> She immediately ordered the application of the new values throughout the company's materials, online and offline, and she branded offices and other premises with quotations and keywords deriving from the values. Departments had to promote the values at every opportunity by explaining them to employees, highlighting the values' importance for the future of the company. Everyone had to sign a document committing to the values. Team-building events were organized specifically for that reason and the whole project was implemented with success and full agreement and support by employees at all levels. A year later the majority of people in the company were nowhere near behaving according to the much-publicized values. There was a competitive attitude among employees, less respect for the work of others, no delegation of responsibility and therefore no emphasis on autonomous work. All in all, very little had changed.

What happened? Why did people, while participating actively in their development and agreeing enthusiastically with the final outcome, not behave as expected? What does it really take to persuade employees to collectively adopt new values, or similar initiatives? These are the main questions the CEO asked her team and external partners to answer for her. The company spent a good deal of money and time in building and communicating the values without any significant results. She wanted to know why. The answer surprised her as well as changed her attitude towards communications forever. Apparently, communicating to people in the company for inducing any type of behavioural change has to be about triggering many brain functions responsible for decision making and behaviour. Targeting only people's rationality and analytical brain centre doesn't work. Effective influence needs a holistic brain approach and brain adaptive leaders need to master it. This is a core BAL skill.

Persuading the brain to act

Throughout this book we have discussed extensively how different brain regions and neural pathways affect the way we think, feel, behave and connect. In doing so, we have touched on the importance of communicating in a certain way in order to affect more powerfully our, and other people's, brains. For example, we explained the significance of symbols and words for creating a desired priming environment in Chapter 6, and we referred to feedback in Chapter 1 as a key strategy for saving brain energy and dealing with ego depletion. The common thread in most of these insights is that the human brain is primarily an information-processing organ (Sullivan, 2015). What it does is to receive information in the form of stimuli from within the human body as well as from the outside world through our senses and neural networks in our bodies, and reacts to these stimulus accordingly. As we have argued, this mainly depends both on the stimuli itself and the way the brain is calibrated, or adapted, to respond. In our view the brain, as an information-processing centre, constantly communicates internally in order to identify the right response to the right stimuli. Communication is a core brain activity, if not *the* core activity. And brain communications for us is all about sending the right message to the right brain structures to persuade the brain to react in a desired way. Since influence is such a central notion of leadership, as rightfully highlighted in the *Harvard Business Review*'s 2013 special issue on influence (Ignatius, 2013), modern leaders are in great need of learning and practising brain-friendly communications in order to effectively persuade others how to, and how not to, behave. A leader without serious persuasion power is like an automobile without wheels: it cannot get very far.

In our quest to identify and implement the most effective brain communications model, for years now we have adopted the triune brain approach, as this was explained in previous chapters. This approach separates the brain and all its structures into three main areas: the old/reptilian brain, the emotional/affective/limbic system brain and the rational/thinking/neocortex brain (MacLean, 1990). Each of these structures requires different information/ stimuli to get them going and a holistic approach is needed to take into account all three in order to achieve any significant behaviour-targeting communications goals. This is the way in which we apply the triune brain approach and it has proven to be very successful in our experience, despite the criticisms that it has attracted in recent years regarding the distinctiveness of its parts in evolution, meaning that the three parts of the brain were not necessarily developed and evolved in such a clear and distinctive way (Pribram, 2002). Hailed as the single most influential idea in neuroscience, since the second world war (Harrington, 1992), we fully agree with the prominent thinker Gerald A Cory (2002) in his assessment that:

> ... despite its current lack of popularity in some quarters of neurobiology, I think that the triune brain concept will continue to be influential, and with appropriate modifications as research progresses, provide an important underpinning for interdisciplinary communication and bridging.

It is the simplicity of the model and its effectiveness in communicating brain functionality to wider audiences, as well as its direct links to cognitive, emotional and behavioural human traits, that makes it a suitable tool to discuss the impact and benefits of neuroscience in leadership and organizations.

Leaders need to target all three brains to persuade and change behaviours. In interpersonal and organizational settings this is successfully captured in Chip and Dan Heath's (2010) model presented in their best-selling book *Switch: How to change things when change is hard*. Without explicitly referring to the triune brain, the Heath brothers expanded on the concept of 'the rider and the elephant', which was first introduced by psychologist Jonathan Haidt (2006). Haidt metaphorically described how our rationally thinking brain functions as a rider – clever and fast but not able to move alone – and our emotional brain functions as an elephant – mobile and powerful, but irrational. Each system needs the other: they have to reinforce one another in order to achieve any meaningful result. These two systems were also described in detail by Nobel Laureate in Economics Daniel Kahneman, in his internationally acclaimed 2011 book *Thinking, Fast and Slow*. The Heath brothers added on a further element called the *path*, highlighting the fact that, besides thinking and feeling, the environment itself can shape our actions – something very close to our discussion of brain automaticity in Chapter 6. The Heaths, based on extensive behavioural experiments as well as on similar arguments and examples developed by other authors and scholars, show that changing behaviours need to follow the recipe below, adapted to the themes of the BAL approach in this book:

1. *Rationalize the direction (or direct the rider for the Heaths)*. The rational brain too often gets stuck in meaningless analysis and endless thought-satisfying debates. It cannot make up its mind easily by itself. To avoid analysis-paralysis, we need clear and simple directions for the whole system to follow. Communicating to this brain is all about clarifying blind spots, utilizing 'bright spots' or success stories, and using simple, direct language. Avoiding information overload and showing the ultimate destination are critical moves in engaging the rider in the most effective way.

2. *Emotionalize motivation (or motivate the elephant)*. The emotional brain is the one that moves the system forward, not the rider. However, an elephant without destination and purpose is as good as a lost elephant. So it needs the rider for instructions. However, instructions alone cannot move the elephant. We move people when we help them adopt a growth mindset, when we break down change in less fear-inducing chunks, and especially when we find the right emotion or mix of emotions suitable for specific teams and individuals. The realization that emotions move people more than ideas, numbers,

orders and procedures is crucial for leaders and managers in adopting the BAL model. We have touched upon this extensively in the book.

3 *Formulate the environment (or shape the path)*. The environment consists of all those aspects that can influence our behaviour and/or our perception. In other words, sometimes the behaviour of someone is the result of neither his/her rational, thinking process nor his/her motivation rooted in an emotion. The behaviour is the result of other external factors that may also have an impact on human behaviour. For example, the physical setting of a manager's office could potentially influence the perception and therefore the behaviour of employees while being in that office. Consequently, our brain automations have to get the right cues from the environment in order to allow the elephant and the rider to move towards the desired destination. Brain automations can either be obstacles or enablers in this path. Habits and the physical environment of the organization as well as the social environment can either hinder or boost our teams' efforts in achieving the behavioural goals we have set.

Unfortunately, most of the corporate initiatives that we have come across around the world pay too much attention to the first element – the rider or the rational brain – and overlook the second and the third. Even when it comes to trying to influence the rider alone, most of the time organizations do exactly the opposite to what modern neuroscience, psychology and behavioural economics all tell us: they overload the rider with, often unnecessary, information and/or they exclude critical information needed to set clearly the destination. For example, in a meeting, managers many at times overemphasize the details or give a lot of information regarding a project or a task, attempting to help employees achieve a result by analysing as much as possible the situation around the task. But doing so, they often miss focusing on key points that are critical for the task completion, misleading and overloading the employees' riders and most probably increase their stress levels. Thus, not only do we not seriously involve the other two parts of the brain but we do not address the rider in the right way either. This is why, as in the opening case to this chapter, we often do whatever we can to persuade our employees to follow the new corporate values, but most people still won't follow them. The rider, the elephant and the path have not been holistically dealt with. Successful behavioural influence requires all of them in a way or another. The faster you adopt this holistic mentality, the better for your leadership persuasion skills.

Measuring and changing the collective corporate brain

The collective brain of any organization, represented by the combined states of the three brains of the employees, is directly related to the culture of the organization. The way people think, feel and behave within their company-specific environment directly affects the success of a corporate culture. Wouldn't it be amazing if there was a way to measure this collective brain in order to better understand its current state and to change it if and when needed? This is exactly what we have been doing in the South East European, Central European and Central African regions since 2011.

One of us (Dr Dimitriadis), in close consultation with the other (Dr Psychogios), has developed a diagnostic tool for measuring and changing the collective corporate brain through internal communications and human resources initiatives. The tool was implemented successfully in multinational and regional companies from many industries, such as retail, distribution, banking and food production and has won a prestigious industry award in the Southeastern region (Dimitriadis, 2014). The tool captures effectively:

A What people know: this part refers to the cognitive brain of the company. It measures the level to which employees are happy with the information they receive concerning both the wider position and future of the company and their own role in, and contribution to, the company (the big and specific pictures, respectively).

B What people feel: this part refers to the emotional/affective brain of the company. It measures the level of a large number of feelings within the organization. We measure all core emotions as well as subjective feelings such as appreciation, support and fatigue.

C What people do: this part refers to the habitual/automatic brain of the company. It measures the level of existence of behavioural habits of employees in their everyday actions such as conflict resolution, intra- and inter-departmental communication, internal and external brand advocacy and others.

The tool, based on employee questionnaires and management interviews, was developed as a response to what we were experiencing in the marketplace: a mismatch between problems and solutions. Companies that had emotional issues attacked with more information while others that lacked well-informed employees replied with emotional statements and visionary speeches. We cannot solve emotional problems with more information just as we cannot solve informational gaps with emotions. Without a clear compass of where the problems exist in the collective brain, we are blind and bound to waste both money and time. On the contrary, by examining the three brains of our company we can establish which one has the actual problem and work on targeted solutions.

The tool has been used to redirect internal communication efforts, to unite various functions on finding real solutions depending on the problem, and in giving the companies' leadership a clear direction for building a winning corporate culture. Dividing brains, business units and departments allows us to identify within the organization, specific pockets of concerns and pockets of brain excellence and use them both appropriately. Maybe one department needs its confidence boosting (emotional treatment) while the other needs new habits of reporting problems (behavioural and procedural treatment). One-size-fits-all does not exist for individual brains and as such it does not exist for collective corporate brains either.

Action box: create your own collective brain diagnosis

You can create your own version of the corporate brain diagnostic tool for your company, department or team. This is how.

A Start by creating three categories: think, feel and do. First, how my people think, what they know and what is on their mind. Second, what my people feel or which are the predominant emotions and which the least present. Third, what are our habits, how do we usually behave to one another or which are the accepted norms of corporate behaviour within and outside the organization.

B Under each category, use the information in this and in other chapters of the book to create a list of variables (or questions) that are relevant to your industry, company or situation. Do not overdo it at the beginning. Ten variables (questions) per category are enough to get you started. Those variables could include:

– For think: how aware are we of the strategic intentions of our organization, do we know of the big issues that bother the company, is it clear how our job contributes to the goals of the company, do we know all the requirements and responsibilities of our job, are channels of communication fully understood?

- For feel: how supportive is the company to its people, is optimism or pessimism the dominant mood, do you feel energized/tired, noticed/ignored, curious/bored, part of a team/isolated and others.
- For do: are we proactive or reactive in problem solving, do we approach others when we are faced with an operational problem that we are responsible for, do we collaborate with other departments effectively, do we recommend the company as an excellent place to work to outsiders and others.

C Interviews (or targeted conversations) is the recommended technique for the first time you implement the tool. Discuss, based on the variables, with at least 10 people from the same department. Note down the basic information for each variable and score each one as high, medium or low depending on their performance. We recommend that organization-wide, quantitative studies should be done with the involvement of experts in such research.

D The outcome is to be able to create first, an overall assessment for the three brains (like thinking brain: high, emotional brain: low, habitual brain: medium) but also a more detailed picture of which variables/elements in each category are high, medium or low. The most problematic areas need your immediate attention. Use the rider, elephant and path methodology to change your people in the areas they scored medium or low. Try to sustain and use as examples the variables that scored high.

Expand the tool to different teams and departments. Then compare and contrast them. Do you see any differences? Are there departmental-specific patterns favouring or limiting any of the variables in the three brains of the company? Act accordingly.

Cialdini's influence

Researching ways of improving your influence skills will inevitably lead you to Robert B Cialdini, the undisputed global guru of persuasion. Cialdini, the Regents' Professor Emeritus of Psychology and Marketing at Arizona State University, published his seminal book *Influence: The psychology of persuasion* in 1984, and since then his six principles of persuasion have become the industry standard. Although he developed these principles mainly for marketing and sales use, they can be easily adopted and applied by leaders within their organization. The principles (Cialdini, 2007) we have been using for years, explained in the context of leadership and management, are:

- *Reciprocity*. Reciprocating an action of goodwill is hardwired in our social brains, as we explained in the previous chapter. Our deep-rooted sense of fairness and moral behaviour asks us to return a good gesture to the people that first behaved kindly to us. Applying Axelrod's recommendation for Chapter 7, a leader needs to start a relationship with goodwill. This can help to open up a positive and potentially perpetual circle of cooperation and at the same time, according to Cialdini, it increases our persuasion power. The gravitational pull of good actions is so strong that it can mould other people's behaviour according to ours. When behaving badly, you can also expect to influence the behaviour of others, this time negatively. In other words, if you personalize others will too, if you smile others will too, and if you innovate others will innovate too.

- *Commitment and consistency*. Much of human behaviour can be attributed to our efforts in decreasing what psychologists call cognitive dissonance. This important concept, first developed by US social psychologist Leon Festinger in the 1950s, suggests that people strive in their lives to achieve internal consistency (Festinger, 1957). This means that we constantly try to match our beliefs with our actions or our many beliefs together. Inconsistency (dissonance) between what we believe and what we did on a certain occasion can cause serious internal conflicts that have to be somehow resolved: either by changing our belief, trying to ignore the problem altogether or behaving consistently soon after (Festinger, 1957). For example, committing your team to a certain course of action will influence them to behave accordingly. In other words, if you want your team to show commitment in a project, you need first to show commitment. But be careful: commitment has to be considered deeply important to them. If not, the internal conflict of not following with their actions will not be that great. Make them believe in the action (being integral to their purpose, for example) and they will do everything to act consistently. Also, you can ask your team to write down the reasons why they will commit to an initiative and to present those to the team. Reasons such as responsibility towards the rest of the team as well as towards end customers that the project targets, reputation of the team and the company, commitment to our values etc could work nicely here and become collective reasons.

- *Social proof.* We are all aware of the effect that a group of people looking up in the sky has on a bystander: the bystander will intuitively also look up. This is social proof, or the principle of people following the social group unconsciously and automatically. The classic and famous Asch conformity experiments demonstrate that people can follow the opinion and behaviour of other people even when these other people are obviously wrong. In the original experiment, 50 participants had to separately go through a series of tests where they had to choose between various options and point to the correct answer. The problem was that in most of these tests, other people present gave obvious wrong replies openly in front of the group. Would the participants follow them or would they state the obvious correct answer (for example: look at these three vertical lines on the paper, which one is the shortest?). Surprisingly enough, only 25 per cent of the participants were not swayed by the others' bad choice, while 75 per cent stated the wrong option (one supported by others) at least once (Asch, 1951). We have conducted a similar experiment with our students many times in the last few years.
We ask all students in a class to read a demanding text on politics or economics. Then we ask them to express their opinion in the class about the main issues in that text. However, the first couple of students to express their opinion have been prepared in advance by us to mislead the conversation by expressing completely and obviously wrong assessments. Then we ask the other students (the real subjects of the experiment) to express their own views. Guess what is happening in most of the cases. The majority of the students, although tending to look somewhat surprised, are influenced by the first two misleading responses and agree with them. Only very few disagree. What seems to be critical here is the timing of the first correct answer in the class. If, for example, it appears straight after the first two misleading ones, the chances the rest will also be correct are higher. If the correct one appears later within the group then it tends to influence less, since more opinions agreeing with the first two misleading ones have emerged prior to the correct one. It is highly intriguing but also quite revealing about our social behaviour, isn't it?

We have social brains and thus we have a tendency to follow the crowd even when it is obviously mistaken. In another favourite experiment of ours, the rate of stealing pieces out of a Petrified Forest park in Arizona increased after a counter-theft campaigned stated:

'Many past visitors have removed the petrified wood from the park, changing the state of the Petrified Forest' (Cialdini et al, 2006). A campaign to decrease theft actually promoted it! So much for common-sense messaging. Our brains tend to follow by default what other people do. When announcing an important meeting, always mention the people that already confirmed participation, especially if they are in high numbers. If participation proves to be limited, do not shout it out. Negative social proof has a strong influential pool, meaning that some people will probably opt out of the next meeting by default if their brains capture the fact that past meetings were ill-attended. Cultural norms within organizations tend to replicate people's accepted behaviour, positive or negative. Pay extra care to which one is which and change it.

- *Authority.* People tend to follow rules by default when those rules are made by figures they deem highly authoritative. The seminal experiments on obedience by Stanley Milgram and Philip Zimbardo mentioned in Chapter 5 prove the point to the extreme. This means that it has to be clear in the company who is responsible for which decisions. Confusion of authority harms your persuasion potential. Furthermore, some classic studies (Patchen, 1974; Pettigrew, 1972; Hickson *et al*, 1971; Mechanic, 1962; French and Raven, 1959) show that expert power can boost the authority effect and increase persuasion. A communicator who is perceived as possessing high expertise can influence both attitudes and the memory of others. By stimulating the caudate nuclei in the brain, an area in the basal ganglia related to feedback processing, learning, expectations of reward, social cooperation and trustworthiness, as well as the medial temporal lobe associated with long-term memory formation and high expertise increase by up to 12 per cent the positive acceptance of a message, meaning attitude, and 10 per cent its recognition, meaning memory (Klucharev *et al*, 2008). These are actually very high numbers compared with older studies (Klucharev *et al*, 2008). Be an expert in your field whatever field this is, meaning strive for excellence in your job area rather than just be good at it. Demonstrate this expertise whenever needed, especially in highly pressured situations, and your influence levels will increase.

- *Liking.* We are more likely to be influenced by the people we like and admire than by those we don't. The three principles of liking and their impact on persuasion are well-evangelized (Dobelli, 2013). First, we like people that are physically attractive. Second, we like people that bear some similarities with us. And third, we like people that like us. This means that leaders and managers have to take care of their physical presence and always be presentable. They also need to present themselves in ways that imply some similarity to their followers. The leader can demonstrate this through statements such as 'I started the same as you', 'we share the same vision' and so on. Lastly, leaders should give compliments whenever possible and appropriate. Part of this persuasion principle includes the claim that we make more effort for people who like us than for people who do not like us. In our experience, this has been consistently proved right. We have observed that people are at their most passionate, proactive and inventive when they are being appreciated.

- *Scarcity*. Our brains have evolved to notice scarcity and to automatically move towards securing vital resources that are about to be in short supply. Prioritizing behaviour that gains access to resources that our perception categorizes as scarce was a necessary evolutionary strategy, since being left out from acquiring those resources could mean failing to survive. Cialdini emphasized the marketing and sales aspect of this principle (as with most of the principles that were originally conceived by him for marketing use), advising marketers to always remind potential customers that the fantastic offer they have is expiring soon or that this product is selling fast and is never to be produced again. In modern organizational environments, leaders should always remind their teams that resources, their current position in the market, even the very existence of the company are not to be taken for granted. Business magazines are filled with cases of once mighty companies who, in the space of only a few years, had to struggle for their survival (such as Nokia, Kodak and others). It is not so much about scaring people. It's all about breaking comfort zones, doing reality checks often and setting the tone for dynamic, proactive and innovative work within the team. Furthermore, creative competitions with short timescales that offer desired prizes to employees can be used to create short boosts of enthusiasm and of new ideas. Hackathons, as they are called, brief and intense project development competitions originating in internet, software, digital and creative companies (Leckart, 2012), work so well because they happen rarely. We believe that if they took place every day they would not have the same effect, plus they would certainly bring the workforce to extreme exhaustion. So, using hackathons, in a relevant form for your company, is highly advisable but as a scarce, meaning not every day, solution.

The six principles of persuasion cannot easily work in all situations and with all people. But we have found that with practice and a bit of reflection you can become skilled in choosing the right principle for the right challenge. Some of them can even become habits of your new BAL leadership approach. Use them and increase your persuasive power.

Talk to the brain

Whenever we discuss the topic of leadership persuasion and the brain with our students and corporate clients we often get the question: 'OK great! But do you have any fast tricks or specific wording that we can apply immediately?' Although we are not supporters of those 'revealing' magic words that create blind followers, there are scientific findings that support the use of specific words or phrases that have immediate influential effects on people. Furthermore, there are conversational styles that can help us increase our persuasion impact and collaborative potential in the long run.

Let's start with those that have a quick-win effect and that we personally use almost daily in our teaching and consulting practices. Psychologist, author and media commentator Rob Yeung did an excellent job in summarizing those fast-track techniques at the end of his book *I is for Influence: The new science of persuasion* (Yeung, 2011). Here they are:

- *'That's not all.'* This makes a good offer even better. As Yeung explains, proposing a new position to someone, presenting an organizational restructuring project, or trying to convince your team to put in double hours this month to boost the numbers can be much more effective if you explain the advantages – then follow them up with even more. Do not play all your cards in your main speech but save surprising benefits for the very end. Using the words '… and that's not all' in the right way has the power to make people feel that they get more than they expected, more than they should or more than they negotiated for.
- *'Because…'.* A very intriguing, and widely cited, experiment showed that if a person is trying to skip the queue for the photocopier, they can obtain a high approval rate if they specify their reason. 'Excuse

me, I have five pages. May I use the Xerox machine because I am in a rush' won a staggering 94 per cent acceptance rate than the simpler 'Excuse me, I have five pages. May I use the Xerox machine', which got 60 per cent (Langer *et al*, 1978). Adding the word 'because' boosts influence drastically. The more serious the situation, the more serious the 'because' has to be. It is a shame how many times managers and leaders just say 'this is not possible', 'not now', 'I will think about it later' or other phrases that create disappointment for their colleagues and subordinates. Just adding 'because' and offering a true and honest reason would leave everyone happier. When asking your team for favours at work or for extra effort, always explain to them clearly and directly why this is needed. You should incorporate this technique in your communication because it will make a huge difference to your effectiveness as an influencer.

- *'I need you to do [something very specific and/or quantifiable].'* Being specific in a request increases manifold the positive impact of that request. Instead of saying 'we have to increase our efforts', it would be better to say 'we have to increase our productivity rate by 7 per cent in comparison to last year in order to catch up with other departments'. Being specific talks effectively to the rider in your brain and sets aside any misunderstandings, confusion or unnecessary pessimisms. If the goal is clear and in small chunks, it is also manageable.

- *'You have a choice.'* Always reminding colleagues, employees and business partners that they actually have a choice in front of important and pressing decisions considerably increases the likelihood of the acceptance of the option you are aiming for. This allows people to feel that they are in control, rather than feeling trapped. A feeling of being trapped or in danger can over-stimulate the amygdala with all the negative consequences previously mentioned (such as stress disorder, fear, anxiety, aggression, etc). We always have a choice in any decision, even in the most difficult ones. Explicitly saying so, particularly with a supporting tone of voice, persuades people that what they will opt for is their own decision. This boosts confidence, engagement and satisfaction. Instead of saying something like 'we have no choice, we need to restructure', it would be better to say 'we have a choice: either we stay as we are and we most probably become irrelevant to our customers within six months or we change now and we beat [the competitors] at their own game!' Can you spot the difference?

Fast persuasion is very handy in our everyday meetings, discussions, announcements and presentations. However, we do not advise becoming a trickster of persuasion. We can rarely completely hide our true capabilities, intentions and leadership potential. In this respect, adopting a conversational style that allows us to become more deeply receptive of others and to respond in a more meaningful and influential way can improve the very core of our brain persuasion capability. This is needed from leaders especially because, as Andrew Newberg and Mark Waldman, authors of the book *Words Can Change Your Brain*, explain (Newberg and Waldman, 2012):

> Although we are born with the gift of language, research shows that we are surprisingly unskilled when it comes to communicating with others. We often choose our words without thought, oblivious of the emotional effects they can have on others. We talk more than we need to.

Newberg and Waldman suggest a solution that they call *compassionate communication* which, when practised in the long term, aligns the two sides of our brains and increases neural resonance between conversing people. Neural reasoning refers to the 'accurate transference of information from one brain to another' (Newberg and Waldman, 2012) boosting the possibility of the right signals being sent and received, and of cooperation emerging. The authors suggest 12 ways by which we can become compassionate communicators, many of them touched upon earlier in this book. These are: adopting a more relaxed attitude, being fully present during a conversation, cultivating inner silence, increasing our positive mentality, reflecting on our deepest values (or purpose), accessing pleasant memories (in order for positive facial expressions to emerge), observing nonverbal cues in others, expressing appreciation, speaking warmly, speaking slowly and speaking briefly, and listening deeply. This list is a very convenient summary for becoming a better, more cooperative, and essentially a more influential converser.

We believe in those and in other similar attributes discussed in this book. Although implementing them all is always easier said than done, we have seen real improvements in ourselves and in others whenever we work closely and seriously on this matter. This starts with the simple but profound realization that communication is hard and most often ineffective if we do not consider how the brain works whenever we aim to have any impact on other people's behaviour. In a very well-received TEDx talk in the University of Strathclyde in Scotland, titled 'The illusion of communication and its brain-based solution', one of us (Dimitriadis, 2015) urged the audience to adopt brain-friendly communications if we are to change this world for the better. Sadly, what we see every day within companies is riders speaking to riders, behaving like there are no other brain parts to be included in decision making, and especially in adopting agreed courses of action. So, when two people are conversing, two frontal lobes agree with each other on what is logical, rational and analytically correct to do, and then two whole brains (not only frontal lobes, which actually become irrelevant after the conversation) just do not move those same two people to act accordingly. Picture this: talking to your colleagues you collectively and swiftly agree that putting 110 per cent of your combined effort behind the new project is crucial for the success of that project. But then it becomes clear that people have not increased their efforts as much as originally agreed upon. Sure, a few of them tried a bit harder but nothing close to what was thought to be agreed. The rider understood but the elephant did not move and the path was not clear. Only when we start considering the brain as a whole, both ours and those of the people we talk to, can we make meaningful decisions that will ultimately lead to behavioural change. Modern leaders need to talk to brains as a whole and not exclusively to the rationality-dominated frontal lobes. Talk to the brain and people will move.

The all-important feedback

In a recent (2014) research work by one of us with colleagues, based on a lot of interviews with managers from different type and size of organizations as well as from different industries and sectors, it was found that managerial feedback is a core cue through which habits in the workplace can change (Psychogios *et al*, 2015). Feedback provides information about work characteristics and attempts to steer performance in a given direction (Fedor *et al*, 2001), and is seen as an integral part of the learning process (Ashby and O'Brien, 2007). In the study it was found that feedback can be used as a way to influence the change of organizational behavioural patterns. Three principles that can be applied when giving feedback were found to be crucial.

First, feedback works better when it is informal. When you want to talk to an employee regarding a behaviour that they repeat which is creating problems, then it is preferable (at least the first time around) to avoid formalized procedures such as asking them to come into your office. Opt for more informal ways. The study showed that managers can achieve better results through informal structures such as talking over a coffee, or walking and talking.

The second principle is that feedback should be as specific as possible. Managers need to directly target, from the beginning of the discussion, the behaviour that needs to be changed. Long and generic discussions should be avoided and it needs to be made clear what the issue of the discussion is and that a decision should be made by the end of it.

The third principle is that the feedback should also demonstrate the benefits that will be achieved after the change of the behaviour. Managers need to discuss these benefits with employees emphasizing what the latter can gain after changing a habitual behaviour that is currently causing harm.

We call the above three principles the *three-fold framework of feedback* in organizations.

It is easy to spot similarities between this study's results and many of the concepts discussed in this and in previous chapters. This is because feedback is a basic function influencing our brains and subsequently our behaviour. It has been found that utilizing carefully informational feedback loops can allow the brain to react fittingly to the initial message and to change our behaviour. Effective feedback loops have four components (Goetz, 2011):

1 *Evidence*. Data regarding a behaviour must be immediately received and presented.

2 *Relevance*. Cold data are turned into meaningful input by way of design, presentation, comparison and context.

3 *Consequence*. A message needs to carry a purpose or wider goal. This answers the question 'why?'

4 *Action*. The individual acts in order to achieve the desired data, thus closing this loop and opening a new one.

As an example, think of a company's effort to improve sustainability-related behaviours at the workplace. If there are recycling targets, the recycling output could be communicated often and with concrete data (evidence), data could always be presented next to the desired target for immediate comparison (relevance), a motivational comment can accompany the data announcement to remind people of the purpose and importance of the activity (consequence), and the loop closes with new actions that hopefully will lead to improved measurements (a new loop is born).

Persuasion stimuli

The recently discovered persuasion powers of brain communication have led to an explosion of *neuroculture*, or the wide diffusion of neuro-scientific insights into society as a whole and into popular culture (Frazzetto and Anker, 2009). We find this process very exciting since it is obvious that we believe that leadership, business, education, politics and life in general will drastically improve if a more brain-based approach is adopted. Effectiveness of communications can increase exponentially leading to substantial savings both in money and time. This is more so in advertising where as the old adage goes: 'Half my advertising is wasted, I just don't know which half' (AdAge, 1999). Neuromarketing – the marketing, advertising and sales sector's effort to develop new, more effective communications based on the neuroscience of persuasion – includes many tips, techniques and models of how companies can achieve more with less. Some of them can be easily applied in a leadership context. Especially, the Six Stimuli model of Patrick Renvoise and Christophe Morin, first published in 2002. This model gives extremely helpful insights of how leaders should communicate directly to the brains of their associates and employees. In the following table you can see the pioneering work of Renvoise and Morin (2007), of which stimuli apply in order to be drastically more effective and efficient in persuasion, in the context of internal communications and management information.

TABLE 8.1 The Renvoise-Morin six stimuli model adapted for leadership

Stimuli	Description	Leadership lesson
Self-centredness	Older and deeper brain structures are primarily concerned with our survival. The individual sense of self-preservation makes our brain notice what refers to us personally.	In meetings, presentations and announcements of important news we always need to consider emphasizing the 'what's in it for me' aspect of our message and to make sure we are taking the receiver's perspective. Use the word 'you' frequently and in a positive and constructive way. And of course there are individual aspects in teamwork although some argue there should not be. But by trying to ignore the individual self within teams and suppress it can be disastrous. Trying to promote it through mutual understanding and reciprocal collaboration can elevate it.

TABLE 8.1 *Continued*

Stimuli	Description	Leadership lesson
Contrast	Our brains have to spend energy in order to spot differences between messages. Thus, they are inclined not to in order to save brain power.	Stop sounding the same in every internal communications campaign and management information delivery. Make it easy for your employees' brains to see what you are saying by always developing messages that are clearly emphasizing distinct features and distinct states. The 'before-and-after' and 'with-and-without' communicational effects apply here and can be used to convince people on new policies and change projects. 'Look at the situation without this new rule and with it. Which one do you prefer?' Such messages utilize contrast effectively for making information easier digested by the brain.
Tangibility	The brain is better at understanding concrete and tangible information than abstract and theoretical information. It spends less energy in capturing the former and much more in dealing with the latter.	Always try to use concrete and clear pictures and words, as well as specific contextual information that make a difficult and abstract topic precise and tangible. Numbers can help, especially if mentioned in a specific context: from 'We will spend a lot on this project' to 'We will spend one million, which is our entire departmental budget, on this project'. The second message has a much stronger impact.
Beginning and end	One of the reasons for story-telling being such an ancient art (think of the Iliad and the Odyssey) is because our brains like to receive information sequentially: with a beginning and an end. So attention is naturally higher in those two points.	Be very careful in choosing your words at the beginning and the end of important announcements. The same goes for when meeting new colleagues and employees: they will clearly remember your first and last words in the meeting if they are powerful. Never give a dull start to a presentation in the hope that later on you will give the juicy details. Attention will be lost long before you reach slide number 18.

TABLE 8.1 *Continued*

Stimuli	Description	Leadership lesson
Visualization	We are visual beings and the brain receives most of its information, and more quickly, through our eyes than other senses. The more graphically visual the message, the better.	Never use blunt, boring or difficult to read or understand materials and slides. The message has to be clear, big and centre-stage. Use fewer but bolder visuals, fewer but bolder words and numbers. Leave minimalist art for the art exhibitions and overburdened Excel slides for MBA students. Following the cultural norms and brand guidelines of your company, create internal campaigns and presentations that are direct and easy to notice and understand. The more striking and clear the visuals, the better the effect. Again, be careful of your organization's unique acceptance reflexes for such efforts.
Emotionality	Brains are emotional organs in the sense that it is emotions that make us move to respond behaviourally to a message. As we saw earlier in this chapter, the elephant moves us and the rider directs.	It is not true that corporate information and communication cannot have emotions. If they couldn't then we would live in an unhealthy, even psychopathic, environment (as explained in Chapter 4). Emotions are an integral, if not central, function of our brains and thus using them correctly can only lead to more impactful communication and persuasion. We have noticed that many managers try to de-emotionalize their messages and thus dangerously lower the impact of what they have to say. By being sensitive to the culture of our organization we can infuse our messages with emotions relevant to what we are trying to achieve in order to maximize the effect. Emotions equal actions. It is as simple as that.

A case in which you would be able to use all these stimuli at the same time would be rare. However, you can use a mix of two or three of them in order to make sure that your message will reach the parts of your audience's brains that will induce behavioural responses. Otherwise you will just talk to the executive part of the brain that might be doing a lot of analysis but not much action.

Keep in mind

Persuasion is a core leadership skill and modern leaders need to be able to influence behaviours in order to achieve their organization's goals. In this direction, we need to 'speak' to all three main functions: thinking, feeling and behaving in order to achieve important results. Frontal lobes, or executive parts, speaking to and agreeing with each other can rarely lead to passionate and meaningful behavioural responses by themselves. Emotions have to motivate and the environment facilitate the desired course of action. Direct the rider (rationality), motivate the elephant (emotions) and tweak the path (habits and procedures) to get maximum results both for individual and collective brains. Use when possible: Cialdini's six principles of persuasion; specific phrases that can fast-track your influence to others (such as 'because'); the 12 points for compassionate conversations; and the six stimuli model of getting a brain's attention. Your increased persuasion power, if used appropriately, will have a significant positive impact both on your, and other people's, work.

References

AdAge (1999) John Wanamaker – special report: the advertising century, *AdAge online*, 29 March, URL: http://adage.com/article/special-report-the-advertising-century/john-wanamaker/140185/, accessed 10 October 2015

Asch, SE (1951) Effects of group pressure on the modification and distortion of judgments, in *Groups, Leadership and Men*, ed H Guetzkow, pp 177–190, Carnegie Press, Pittsburgh

Ashby, FG and O'Brien, JRB (2007) The effects of positive versus negative feedback on information-integration category learning, *Perception and Psychophysics*, 69, pp 865–878

Cialdini, RB (2007) *Influence: The psychology of persuasion*, Revised Edition, HarperCollins, New York

Cialdini, RB, Demaine, LJ, Sagarin, BJ, Barrett, DW, Rhoads, K and Winter, PL (2006) Managing social norms for persuasive impact, *Social Influence*, 1 (1), pp 3–15

Cory, GA (2002) Reappraising MacLean's triune brain concept, in *The Evolutionary Neuroethology of Paul MacLean: Convergences and frontiers*, eds GA Cory, and R Gardner, pp 9–27, Greenwood Publishing Group, Westport

Dimitriadis, N (2014) Neuromarketing is the future, eKapija Business Portal-SEE Region, 18 February, URL: www.ekapija.com/website/en/page/842820/Nikolaos-Dimitriadis-CEO-of-DNA-communications-Neuro-marketing-is-the-future, accessed 16 October 2015

Dimitriadis, N (2015) The illusion of communication and its brain-based solution, TEDx The University of Strathclyde, Glasgow, 25 April

Dobelli, R (2013) *The Art of Thinking Clearly*, Spectre, London

Fedor, DB, Davis, WD, Maslyn, JM and Mathieson, K (2001) Performance improvement efforts in response to negative feedback: The role of source power and recipient self-esteem, *Journal of Management*, 27, pp 79–97

Frazzetto, G and Anker, S (2009) Neuroculture, *Nature Reviews Neuroscience*, 10 (11), pp 815–821

French, JRP and Raven, BH (1959) The bases of social power, in *Studies of Social Power*, ed D Cartwright, pp 259–269, Institute for Social Research, Ann Arbor

Goetz, T (2011) The feedback loop, *Wired Magazine*, July, pp 126–133, 162

Haidt, J (2006) *The Happiness Hypothesis: Finding modern truth in ancient wisdom*, Basic Books, New York

Harrington, A (1992) At the intersection of knowledge and values: Fragments of a dialogue in Woods Hole, Massachusetts, August 1990, in *So Human a Brain: Knowledge and values in the neuroscience*, ed A Harrington, pp 247–324, Springer Science and Business Media, New York

Heath, C and Heath, D (2010) *Switch: How to change things when change is hard*, Crown Business, New York

Hickson, DJ, Hinings, CR, Lee, CA, Schneck, RS and Pennings, JM (1971) A strategic contingencies theory of intra-organizational power, *Administrative Science Quarterly*, 16, pp 216–229

Ignatius, A (2013) Influence and leadership – editorial on the HBR Special Issue on Influence: How to get it, how to use it, *Harvard Business Review Online*, URL: https://hbr.org/2013/07/influence-and-leadership, accessed 20 September 2015

Klucharev, V, Smidts, A and Fernández, G (2008) Brain mechanisms of persuasion: How 'expert power' modulates memory and attitudes, *Social Cognitive and Affective Neuroscience*, 3 (4), pp 353–366

Langer, EJ, Blank, A and Chanowitz, B (1978) The mindlessness of ostensibly thoughtful action: The role of 'placebic' information in interpersonal interaction, *Journal of Personality and Social Psychology*, 36 (6), pp 635–642

Leckart, S (2012) The hackathon is on: pitching and programming the next killer app, *Wired online*, 17 February 2012, URL: www.wired.com/2012/02/ff_hackathons/all/1, accessed 17 October 2015

MacLean, PD (1990) *The Triune Brain in Evolution: Role in paleocerebral functions*, Plenum Press, New York

Mechanic, D (1962) Sources of power of lower participants in complex organizations, *Administrative Science Quarterly*, 7, pp 349–364

Newberg, A and Waldman, MR (2012) *Words Can Change Your Brain: 12 conversational strategies to build trust, resolve conflict, and increase intimacy*, Hudson Street Press, New York

Patchen, M (1974) The locus and basis of influence on organizational decisions, *Organizational Behavior and Human Performance*, **11**, pp 195–221

Pettigrew, A (1972) Information control as a power resource, *Sociology*, **6**, pp 187–204

Pribram, KH (2002) Pribram and MacLean in perspective, in *The Evolutionary Neuroethology of Paul MacLean: Convergences and Frontiers*, eds GA Cory and R Gardner, Greenwood Publishing Group, Westport, pp 1–8

Psychogios, A, Szamosi, L, O'Regan, N and Blackory, F (2015) Can feedback alter organizational routines? Five-fold dimensions of managers' feedback in changing organisational routines, *Hull University Business School Working Paper Series*, The University of Hull

Renvoise, P and Morin, C (2007) *Neuromarketing: Understanding the 'buy buttons' in your customer's brain*, SalesBrain LLC, San Francisco

Sullivan, J (2015) Born to trust: The brain evolution as a social organism – a conversation with Louis Cozolino, PHD, *BrainWorld*, **6** (2), Winter, pp 50–53

Yeung, R (2011) *I is for Influence: The new science of persuasion*, Macmillan, London

SUMMARY OF PILLAR 4: RELATIONS

TABLE S.4 Summary of pillar 4

Focus on collective effort	Focus on collaboration rather than on individualistic effort.
Connections matter	Do not neglect connections with others that have a dramatic impact on brain functions, and can have an impact on leadership performance.
Understand how others are thinking	Engage directly with others by creating a comfortable situation, being as clear as possible, listening actively.
Be aware of imitation	Leaders should be aware of imitation when shaping other people's moods, attitudes and behaviour.
Persuade	Persuade by adopting the triune brain approach: • rationalize the direction; • emotionalize motivation; • formulate the environment.
Evaluate the collective brain	Use the three-fold diagnostic tool for evaluating the collective brain: • what people know; • what people feel; • what people do.
Be aware of influence	Be aware of the following principles of influence: • reciprocity; • commitment and consistency; • social proof; • authority; • liking; • scarcity.

TABLE S.4 *Continued*

Talk to the brain	There are specific words or phrases that have immediate impact: • *'That's not all.'* • *'Because...'* • *'I need you to...'* • *'You have a choice.'*
Be aware of feedback	Emphasize: 1 informal feedback; 2 specific feedback; 3 benefits-oriented feedback.
Persuasion stimuli	Six specific stimuli can make leaders more effective in persuasion: 1 self-centredness; 2 contrast; 3 tangibility; 4 beginning and end; 5 visualization; 6 emotionality.

CONCLUDING REMARKS
The future, the brain and the BAL approach

The very fact you are reading this sentence means that our application of the BAL model in writing the book was successful. Whether the book itself is successful depends on the results you get by applying the BAL model. You succeed, we succeed.

But this was not the first time we applied the BAL model. Our involvement with neuroscience, psychology, sociology, anthropology, behavioural economics and, of course, with the sciences of management and marketing, has inevitably calibrated our brains to apply many of the principles of the book almost automatically. For the past 10 years both of us have engaged in multiple roles in our professional lives: lecturers, speakers, trainers, researchers, mentors, coaches, consultants, managers and entrepreneurs. These roles helped us generate multiple experiences with the BAL model, both directly and through the people we taught, advised and coached. Thus, we have not included in the book any insight or recommendation that we have not tried ourselves and that did not produce tangible results. It is this deep-rooted, experiential and shared belief we have in the BAL approach that gave us the incentive to write the book and make it accessible to more people around the world. Our promise is this: by applying the BAL principles your brain will become your strongest ally in leadership.

Ground-breaking discoveries about the brain have just started to surface. What we currently know is probably just the tip of the iceberg. As Gary Marcus and Jeremy Freeman, editors of the book *The Future of the Brain: Essays by the world's leading neuroscientists* declared, 'There's never been a more exciting moment in neuroscience than now' (Marcus and Freeman, 2015). We are expecting that the next two decades will be seminal in revealing even more intriguing insights and facts about how our neurons fire, interact and impact on our behaviour. We did our best, within the usual constraints, to include in the book the latest findings and discussions about the brain, combining them with classic studies that go back to the middle of the 20th century. We feel that this blend of classic and modern (with an emphasis on the latter), based on extensive research conducted in brain-related literature, together with the tangible results we have observed, gives the BAL approach its edge. We believe that the four pillars will remain as they are for a long time to come. Keep an eye out though, and an open mind, for new findings, new research, new applications of neuroscience and behavioural sciences. Use those new findings to develop it even further yourself! Your brain is plastic and it loves to grow. Use the BAL approach to do so.

From all the new developments in the wider field of neuroscience, the one that will have a profound impact on the way we perceive and deal with the brain, is brain-interacting or brain-altering technology. Initial findings, in what has been termed *neurotechnology* (Donoghue, 2015), are staggering, and create huge hopes for people with brain damage. BrainGate (www.braingate2.org/) and other similar scientific initiatives around the world are building devices and computer software that interact directly with neurons in the brain. It seems that not only naturally occurring neuroplasticity can improve our brains, but also electrodes implanted in our heads and sensors on the top of our skulls. Technology will be able to bypass our executive brains. It will reach deeper structures to evoke direct responses. It will acquire clearer information of thoughts and feelings in our brain. And it will help two or more brains to directly interact and communicate. Here is a selection of highly intriguing studies on the topic:

1 *Direct brain-to-brain communication.* A team led by Rajesh PN Rao, Director of the NSF Center for Sensorimotor Neural Engineering and Professor of Computer Science and Engineering at the University of Washington, conducted the first ever brain-to-brain interface in August 2013 (Rao *et al*, 2014). In the experiment, two people sitting in different locations had to cooperate with each other in order to win a computer game. The first, called the sender, was looking at the screen of the game thinking what move he and his game partner should do, but did not have access to the physical controls. The second, called the receiver, made the moves in the game by using his hand but without having access to the screen to see for himself which move was required. So he depended on the sender. The sender was wearing a brainwaves-reading device (EEG) that picked up his thoughts on making the right moves to win the game. The receiver was also wearing a device, called transcranial magnetic stimulation (TMS) that could make him move his hand involuntarily by stimulating magnetically specific motor centres in his brain. They managed to win the game. The thoughts of the sender moved the hand of the receiver in the right direction to play, according to the simple rules of the video game.

2 *Image reconstruction.* In a paper published in 2011, Sinji Nishimoto and associates announced their success in capturing images from people's minds watching nature films, and reconstructing them again on a separate screen. People in an fMRI scan were watching these films while, at the same time, the scientists were recording what was happening in those people's brain areas associated with visual imaging, specifically at the occipito-temporal visual cortex. By using advanced models of visual representation, they managed to reconstruct the dynamic images viewed by participants to a satisfactory degree (Nishimoto *et al*, 2011). Although somewhat fuzzy, these reconstructed images were good enough to create high hopes for improvement in the near future. As Thomas Naselaris, one of the authors of the study, stated more recently: 'The potential to do something like mind reading is going to be available sooner rather than later. It's going to be possible within our lifetimes' (Requarth, 2015).

3 *Sharing thoughts.* A study published in 2014 demonstrated that two brains connected could share a thought between them through non-invasive neurotechnologies (Grau *et al*, 2014). A person sitting in Kerala in India (the emitter) was asked to imagine that he was moving either his hand or foot. The brain-computer interface device he was wearing on his head captured the signal in his brain, generated by his thought, and translated it into a simple binary code (0 and 1) that represented each imagined action. Then the researchers assigned a word to each imagined action, the Italian word *Ciao* and the Spanish word *Hola*, depending on if he thought of moving his hand or his foot. The emitter knew this so he could decide which word it would be when he made a move. The generated binary codes were then emailed to France, where another person (the receiver) was sitting in Strasburg with his eyes closed and wearing another device on his head. The device translated the binary code it received into flashes of light that the receiver saw in his mind's eye. Knowing in advance what the flashes could mean, he could say the words *Ciao* or *Hola*, thus replicating directly what the emitter had decided to send. In essence, the two brains communicated directly with each other over the internet. The emitter intended to say either *Ciao* or *Hola* and the receiver knew which one. Giulio Ruffini, one of the authors of the study, said very openly that:

> ... you can look at this experiment in two ways. On the one hand it's quite technical and a very humble proof of concept. On the other hand, this was the first time it was done, so it was a little bit of a historical moment I suppose, and it was pretty exciting. (Eveleth, 2015)

With such leaps in brain understanding and interaction, the renowned physicist and futurologist Michio Kaku rightfully included in his 2015 book *The Future of the Mind*, topics as controversial and mystical as telepathy, telekinesis, mind control, silicon consciousness and mind beyond matter. In his own words, from the opening lines of the book (Kaku, 2015):

> The two greatest mysteries in all of nature are the mind and the universe... If you want to appreciate the majesty of the universe, just turn your gaze to the heavens at night, ablaze with billions of stars... To witness the mystery of our mind, all we have to do is stare at ourselves in the mirror and wonder, What lurks behind our eyes?... But [until recently] the basic tools of neuroscience did not provide a systematic way of analyzing the brain.

Now, it does.

What do all these advances mean for brain-based leadership? Although it is difficult to say, we can envision a few things. Firstly, leaders will take brain training methods and initiatives, like the BAL approach, more seriously – both for themselves and for their organizations. This also means, hopefully, that brain-friendly approaches will be introduced early on in education in order for children and teenagers to use brain-enhancing approaches to deal more effectively with the challenges they face at school and in life.

Secondly, leaders and managers will be using mobile applications, computer programs and various devices to help them get neuro-feedback (how their neurons behave triggered by controlled stimuli) for many of their daily tasks. Already feasible examples include:

- recruiting personnel involving facial recognition and EEG (electroencephalogram) tests;
- detecting brainwaves of people at the office in order to enable them to get faster into their flow state and thus, to be more productive and fulfilled; and
- mapping emotions through voice recognition and other methods during meetings to help companies create a more collaborative and exciting environment for their employees.

At the same time, advanced data analytics from all possible digital sources will push the boundaries in discovering deeply rooted behavioural patterns at work.

Concluding remarks

Thirdly, further in the future, we see the direct implementation of neurotechnology and brain-to-brain communications in leadership, probably starting from global corporations, governments and even education. Of course, all these are predictions and, as such, they need to be considered with caution. More so, since many of them will not become reality if the two important issues of privacy and ethics are not adequately resolved.

Of course, applied neuroscience and neurotechnology are like any other human tool: they can be used with good or bad intentions. For example, it can be used for promoting egoistic goals and for manipulation, or for creating shared and collective benefits. It can also be used as a means for conflict resolution or as a means of conflict expansion, and even for war. A special report in the magazine *Foreign Policy* revealed that neuroweapons are already in the making and that considerable amounts of money are being poured from defence agencies and companies into integrating neurotechnology with humans and infrastructure systems (Requarth, 2015). 'Welcome to the Neurowars' the front page of the September/October 2015 issue declared so, unfortunately, we might not be so far from such an undesirable future.

Brain structure, company structure

Organizational structures will not be immune to neuroscience. The way we organize jobs, tasks, authority levels and communication lines within organizations will follow the way our brains are organized. Internal brain structures will be mirrored in team, departmental and organizational design and charts. Does it sound too farfetched? It is already happening. Renowned leadership guru John P Kotter explained in his seminal *Harvard Business Review* article 'Accelerate' in the November 2012 issue, and expanded on in his book *XLR8: Building strategic agility for a faster moving world*, his findings of a dual operating system within corporations. This dual system consists of the traditional, pyramid-like, formal and inflexible hierarchy, with more informal, spontaneous, flexible, startup-like and project-driven networks (Kotter, 2014). The first system needs to exist to ensure continuity of delivery and stability while the second is vital for boosting innovation, creative disruption and future competitiveness. Companies need both as our brains need both: rationality and emotions need to work together, in harmony. If not, brains get dysfunctional and companies get outpaced by leaner and more creative startups.

Marketing experts Marc de Swaan Arons, Frank van den Driest and Keith Weed published their article 'The ultimate marketing machine' in the

Concluding remarks

> July/August issue of the *Harvard Business Review*, an issue dedicated to *Think, Feel, Do: The new basics of marketing*. In that article, the authors described the optimum marketing function structure as one mirroring key brain functions (Arons *et al*, 2014). They suggested that in order for marketing to work better, 'think' tasks (including research and analytics), 'feel' tasks (including customer engagement management and media relations) and 'do' tasks (including content creation and production) need to be handled but by different teams, each one staffed with specialists in their respective brain function. Mildly put, this is a revolution in the making. Leaders need to urgently become neuroscience-ready and lead this revolution into new realms. Those who do will be the winners.

The main point here is that all pillars of the BAL approach will almost certainly improve and expand in the future. Let's see briefly how a few of the issues discussed earlier will help (with some already happening):

- *Pillar 1*. Thinking will become smarter. We will spot biases, false patterns and ego depletion faster than ever due to self-assessment applications and better analysis of data from the world around us. We will ask better questions, place smarter strategic bets and understand sooner whether we are pursuing our purpose as an organization, since communication between people will become more brain-friendly.
- *Pillar 2*. Feelings will become more meaningful. Our capacity for emotional intelligence will increase due to feedback technology that will allow us to understand better what we and others feel at work. Also, emotional styles and moods will be tracked and managed better through data analytics and emotional-detecting interfaces.
- *Pillar 3*. Brain automations will work more to our advantage. Neurodesign will help us create better spaces, both online and offline. Those spaces will prime our brains for better cooperation, more creativity and higher productivity. We will also be able to track habits more easily through apps and sensors in mobile and other devices.
- *Pillar 4*. Relations will become stronger and more motivating. Social media are already revolutionizing the way people connect, work and live, inside and outside organizations. Building useful bridges will be easier than ever but we will also be able to nurture and evaluate those bridges more efficiently. Persuasion will be digitally

aided through advanced analytics and even computer-generated information and suggestions. This will be largely based on what is called predictive neuroscience, which is evaluating, modelling and predicting neural behaviour in our brain (Markram, 2013). At the same time, internal brain structures will form the basis for corporate restructuring approaches worldwide. Neurotechnology will optimize communication between parts in these new structures in order to become super-efficient and super-effective, as our brain already is.

Complexity asks for great leaders: great leaders in any social or economic activity and in any company or institution; great leaders from any age group and from any place on earth. Complexity also requires phronetic-leaders. Since past knowledge is not enough to deal with the future complex problems, we need to cultivate more of our practical, day-to-day judgements in making decisions and in behaving in a social context. We aspire that the BAL approach will be a significant step towards strengthening our 'phronesis' and therefore our individual and collective leadership potential, and unleashing it into the world. Go on and unleash it!

Keep in mind

... all the *Keep in mind* sections in this book. We strongly recommend that you revisit them often to refresh your memory and to boost your brain leadership into new realms!

References

Arons, MDS, van den Driest, F and Weed, K (2014) The ultimate marketing machine, *Harvard Business Review*, July–August, 92 (7), pp 54–63

Donoghue, J (2015) Neurotechnology, in *The Future of the Brain: Essays by the world's leading neuroscientists*, ed G Marcus and J Freeman, pp 219–233. Princeton University Press, Princeton

Eveleth, R (2015) I emailed a message between two brains, *BBC Futureonline*, URL: www.bbc.com/future/story/20150106-the-first-brain-to-brain-emails, accessed 8 October 2015

Grau, C, Ginhoux, R, Riera, A, Nguyen, TL, Chauvat, H, Berg, M, Amengual, JL, Pascual-Leone, A and Ruffini, R (2014) Conscious brain-to-brain communication in humans using non-invasive technologies, *PLoS ONE*, 9 (8), e105225. doi:10.1371/journal.pone.0105225

Kaku, M (2015) *The Future of the Mind: The scientific quest to understand, enhance, and empower the mind*, Anchor Books, New York

Kotter, JP (2014) *XLR8: Building strategic agility for a faster-moving world*, Harvard Business Review Press, Boston

Marcus, G and Freeman, J (2015) Preface, in *The Future of the Brain: Essays by the world's leading neuroscientists*, ed G Marcus and J Freeman, pp xi–xiii. Princeton University Press, Princeton

Markram, H (2013) Seven challenges for neuroscience, *Functional Neurology*, 28 (3), pp 145–151

Nishimoto, S, Vu, AT, Naselaris, T, Benjamini, Y, Yu, B and Gallant, JL (2011) Reconstructing visual experiences from brain activity evoked by natural movies, *Current Biology*, 21 (19), pp 1641–1646

Rao, RPN, Stocco, A, Bryan, M, Sarma, D, Youngquist, TM and Wu, J (2014) A direct brain-to-brain interface in humans, *PLoS ONE*, 9 (11), e111332. doi:10.1371/journal.pone.0111332

Requarth, T (2015) Mind field, *Foreign Policy*, September/October, 214, pp 50–59

EPILOGUE

The BAL approach has become integral to the way that both of us experience our professional and personal lives. Writing this book was no exception. It required us to put into practice all four pillars of the BAL approach carefully, consistently and creatively. This is how.

On thinking, we had to use our willpower carefully since we both worked full time while writing it. We had to keep away as much as possible from the burnout effect, from various cognitive biases, and from misleading pattern recognition, especially while deciding which relevant literature to use and which suggestions and examples. We had to constantly ask the right questions for the development of the book and remind ourselves of the higher purpose for writing it. We had to be creative in solving various challenges, get into the flow as often as possible (for fast progress) and to overstretch our memory to make sure that we didn't leave out anything we deemed important. Enhancing our growth mindsets was absolutely necessary!

On emotions, the majority of emotions experienced while writing this book ranged, thankfully, from positive to extremely positive. So, the dominant mood was that of *The Rocket*. However, any negative emotions that arose during any mishaps were also put to good use: they made us more determined and more resilient towards our aim. All these emotions were what kept us going (as emotions do). The dominant core emotion was *excitement*, or *seeking*, and the emotional equation we experienced the most was *curiosity* (= wonder + awe).

On brain automations, we took advantage of priming by making sure that our desks were always full of materials related to the book. We made daily online meetings between us a routine, wherever in the world we happened to be. We impulsively bought any book or magazine on the market (usually in airport bookstores throughout Europe and Asia) that mentioned neuroscience or the brain, as well as reading any latest special report on the subject in various online sources. These are also habits we've shared for the best part of 10 years.

On relations, we did our best to stay closely connected to a number of people, including each other, our publisher's wonderful team, professionals and scientists with similar interests around the globe and, of course, our students, colleagues and clients. Weak links proved as important as strong ones in terms of new ideas, problem solving and reaching out to the world.

Most importantly, we never stopped trying to positively influence a variety of people, such as (again) each other, our business and life partners, and our support team at our publisher in order to finish the book in the best way possible. At the same time, by mentioning the book to people we met in class, companies and events, we tried to sway them positively and create anticipation.

Thank you for reading this book and, hopefully, for sharing our view on this new, brain-based, era of leadership. There are now two clear paths ahead of you: either you continue having the same thoughts, emotions, habits and relations as you did before reading it OR you apply the four pillars and you become an even better version of your leadership self. The choice is yours.

INDEX

Achor, Simon   118–19
active listening   99
adaptability
    in an unpredictable world   70–73
    requirement in a complex world   6
adaptive leadership style   7
adaptive unconscious   134
amygdala   85
    and Theory of Mind   173
    and trust   186–87
amygdala hijacking   47–48, 88
analysis-paralysis   135
anti-fragility   71
Apple   55
Aristotle   34
Arons, Marc de Swaan   223–24
attention (focus)   92–93
Attention Schema Theory   170
autocratic-controlling leadership style   7
automatic responses *see* brain automations
avoid/approach systems   105, 108
Axelrod, Robert   167–68

Babiak, Paul   84
BAL *see* brain adaptive leadership
Bargh, JA   139
Bear, Meg   188
Bechara, Antoine   67
behavioural economics   3, 4
being in the zone *see* flow state
Berger, Warren   41
Berka, Chris   60
betting, purposeful betting   71–73
bias, cognitive bias in challenging situations   39–41
Biswas-Diener, Robert   122
black swans (unpredicted events)   71
Blake, Frank   42–43
bliss leadership   116–21
Boyatzis, Richard   177–78
Brafman, Ori   184
Brafman, Rom   184
brain
    adaptability   6
    attraction of pattern recognition   36–37
    basic emotions   105–11
    capability of lifelong learning   54
    changes throughout life   52–54

evolutionary development   34–36
focus of leadership research   3–4
hypofrontality   63
influence of the older brain structures   34–36, 38
influences on behaviour   3–4
insights from neuroscience   3–4
mirror neurons   177–78
neuroplasticity   6, 52–54
socially wired brain   169–72
survival as priority   34–36
survival protocols   46–48
brain adaptive leadership (BAL)   3
    adaptive element   4–6
    background to the BAL approach   11–12
    brain element   3–4
    experiences of application   219
    future developments   219–25
    leadership element   6–7
    Pillar 1 (thinking)   8–9, 15–74, 76–77
    Pillar 2 (emotions)   9, 79–124, 127–28
    Pillar 3 (brain automations)   9–10, 131–55, 160
    Pillar 4 (relations)   10–11, 163–217, 216–17
    summary of the approach   8–11
    writing this book   227–28
brain automations (Pillar 3)   9–10, 131–55, 160
    adaptive unconscious   134
    and expertise   152–54
    effects of technology   148–49
    embodied cognition   149–52
    gut feelings   134, 152
    habits   143–48
    influence of the physical environment   148–52
    intelligence of the unconscious   134
    interactions with the body   149–52
    intuition   134, 152–54
    power of the unconscious mind   131–36
    priming   136–43
    readiness potential   133–35, 136–37
    summary of Pillar 3   9–10, 160
brain chemicals, role in trust   187–88
brain hijacking   47–48
brain-interacting/altering technology   220–25

Index

brain misfires 39–41
brain power 17–31
   and strong values 21–23
   avoiding burnout syndrome 23–26
   benefit of immediate feedback 23
   contribution of higher-level thinking 21
   distinguishing tasks from leadership 31
   energy requirements of the brain 18–19
   energy used for self-control 19–20
   how the brain uses its resources 18–19
   leadership willpower muscle 19–23
   performance impact of multitasking 26–30
   pressures of the business world 17–18
BrainGate initiative 220
brainwaves associated with flow state 60
Brooks, Alison Wood 96
burnout syndrome 23–26
business, application of insights on emotions 108–11
Byrne, Mike 63

Cabane, Olivia Fox 185
carrot and stick approach to management 103–05
chaos theory 5
Christakis, Nicholas 181–83
Chrysikou, Evangelia 63
Cialdini, Robert B 200–05
clear your mind 33–49
   amygdala hijacking 47–48
   brain hijacking 47–48
   brain misfires 39–41
   brain survival protocols 46–48
   dangers of groupthink 37–38
   dealing with negative thoughts 46–48
   influence of the older brain structures 34–36, 38
   pattern recognition traps 33–34, 36–37
   power of asking questions 41–43
   relaxation techniques 48
   risk of cognitive bias 39–41
   style of questioning 43–46
   when the ape takes control 46–48
cognitive bias in challenging situations 39–41
collective corporate brain, measuring and changing 198–200
Colvin, Geoff 188
communication 193–213
   brain-friendly communications 205–08
   changing behaviours 193–200
   compassionate communication 207–08
   importance of feedback 208–09
   influential words and phrases 205–08
   persuading the brain to act 194–200
   six principles of influence 200–05
complexity
   characteristic of human society 4–5
   defining 4
   in social sciences 5
complexity theory 5
complication, distinction from complexity 4
Conley, Chip 85, 114–16
context sensitivity 91–92
cooperation 163–72
   and the Prisoner's Dilemma 164–69
   socially wired brain 169–72
corporate culture
   power of fear 103–04, 110
   priming effects 140
   situations favouring the biased mind 39–41
corporate values, changes in challenging situations 39–41
Cory, Gerald A 195
Covey, Stephen 146
Cozolino, Louis 170–71
creativity 62–66
   allowing ideas to 'bake' 63, 64
   and brain activity 63
   and neuroplasticity 66
   bias against in business 64
   role in business 62–63, 64
   strategies to foster 63, 64–66
Csikszentmihalyi, Mihaly 60
curiosity, and creativity 62

Damasio, Antonio 83, 89
Davidson, Richard 85–93
Dawkins, Richard 166
de Groot, Adriaan 153
decision making
   and good memory 67
   dangers of groupthink 37–38
   readiness potential in the brain 133–35, 136–37
Descartes, René 34
Dijksterhuis, Ap 134–35
dopamine 107, 144
   and pattern recognition 36
   and reward 87
Dostoyevsky, Fyodor 135
downer mood 95, 96
Doyen, S 139
Duhigg, Charles 145
Dweck, Dr Carol S 73

ego depletion, effect on willpower   20
Ekman, Dr Paul   87–88
  model of emotions   106–07
embodied cognition concept   149–52
emotional agility, ways to develop   120–21
emotional dimensions of leadership   99–100
emotional intelligence   97–100
  relationship management (external management)   99
  self-awareness (internal awareness)   98
  self-management (internal management)   98
  social competence (external awareness)   98
emotional opposites   112
emotional quotient (EQ), as an empowerment qualification   97–100
emotional styles   85–93
  attention (focus)   92–93
  context sensitivity   91–92
  outlook   86–87
  pessimism–optimism continuum   86–87
  resilience   86
  self-awareness   88–91
  social intuition   87–88
  teams   93
emotions (Pillar 2)   9, 79–124, 127–28
  3D+3L approach to understanding   90–91
  and memory   68–69
  and morality   84
  and motivation   84–85
  application of insights in organizations   108–11
  basic emotions   105–11
  classification   85
  combinations of emotions   111–16
  dangers of submission and obedience   113
  dealing with core emotions   108–11
  decoding   89–91
  distinction from feelings   89
  effects of lack of   83–84
  emotion-run brain   83–85
  exclusion in management training   84
  link to neurotransmitters   107
  re-evaluation of beliefs about   121–24
  role in business   81–100
  role in decision making   81–100
  summary of Pillar 2   9, 127–28
empathy   84, 99, 175, 188
energy requirements of the brain   18–19
environment, emotional synchronization with   91–92
Epley, Professor Nicholas   175

exhaustion, and burnout syndrome   23–26
expertise, and automaticity   152–54
extrinsic motivation   55–56, 57–58

fear, power in the workplace   103–04, 110, 112–13
feedback
  developing willpower   23
  importance in communication   208–09
feelings, distinction from emotions   89
Flow Genome Project   60
flow state
  and optimal performance   59–62
  brainwaves associated with   60
  conditions for achieving   61
Foer, Joshua   67, 69
Fowler, James   181–83
Fox, Elaine   53, 85, 86, 105
Freeman, Jeremy   219
Freud, Sigmund   133

Gage, Phineas   83, 84
Gallate, Jason   93
Gigerenzer, Gerd   134
Gladwell, Malcolm   134
Goleman, Daniel   97, 98, 177–78
Gonzales, Maria   90
Granovetter, Mark   179–80
Graziano, Michael   170
groupthink   37–38
growth mindset   73
guru mood   95, 96
gut feelings   134, 152

habits   143–58
  and goals   144, 145
  changing an unwanted habit   145–48
  forming a new habit   144
  habit cycle   145
  leadership habits   146–48
  organizational routines   144–45
Halligan, Professor Peter   170
Hamann, Stephan   69
Hammond, Claudia   116
happiness
  benefits in the workplace   118–19
  relationship to success   118–19
  study of   116–21
  ways to create and boost   120–21
Hare, Robert   84
Hattula, Johannes D   175
Heath, Chip   196
Heath, Dan   196
Heidt, Jonathan   196

Index

higher-level thinking, contribution to brain power 21
hippocampus 67–68, 91–92
Home Depot 42–43
Hood, Bruce 54
human connectivity 178–86
hygiene and motivation theory (Herzberg) 117

Iacoboni, Marco 177
imitation 177–78
infocracy paradigm of social organization 5–6
information, power of 5–6
informed enquiry 41
innovation *see* creativity
Inside Out (movie, Pixar Studios) 123
intrinsic motivation 55–58, 99
intuition 134, 152

Janis, Irving 38
Johannson, Frans 71–73

Kahneman, Daniel 139, 196
Kaku, Michio 222
Kashdan, Todd 122
King, Martin Luther 55
Klein, Gary 153–54
Kotler, Steven 60
Kotter, John P 223

laser focus 93
leadership
    and purpose 56–57
    as an attitude 1–3
    bliss leadership 116–21
    distinguishing from tasks 31
    element of BAL 6–7
    emotional dimensions 99–100
    human connectivity 178–86
    insights from neuroscience 2–3
    practical judgement (*phronesis*) aspect 7
    psychology of following a leader 6–7
    six stimuli model of persuasion 210–13
    understanding what makes a great leader 1–3
leadership habits 146–48
leadership network connections matrix 180–81
leadership styles 7
Libert, Benjamin 133
Lieberman, Professor Matthew D 174–75
lifelong learning 54
limbic system 35, 47
Lobel, Thalma 151

logos, priming effect 141
long-term memory 67–70
love, emotional components 114
Lovheim, Hugo 107–08
low-level thinking 21

machines replacing humans at work 188–89
Mackay, Professor Donald G 68
management, carrot and stick approach 103–05
management education, simplistic view of emotions 103–05, 108
Marcus, Gary 219
Marean, Curtis 171
memory
    and emotions 68–69
    benefits of a good memory 67–70
    importance for leaders 67
    minimizing memory loss 68
    role of the hippocampus 67–68
    short-term memory 67
    strengthening long-term memory 67–70
    utilizing associations to help remember 69–70
Milgram, Stanley 113, 203
Miller, Dr Liz 94, 95–96
mirror neurons 177–78
Mischel, Walter 19
mismatch, and burnout 25–26
mood 94–97
    and emotions 94
    benefits of positive mood 94–97
    definition 85
    Miller's classification 95–96
morality 84
Morin, Christophe 210–12
motivation
    and emotion 84–58
    and purpose 54–59
motivational-engagement leadership style 7
Mourkogiannis, Nikos 56
Mueller, Jennifer 63–64
multitasking
    dangers of high cognitive load 26–30
    impact on performance 26–30

Nadler, Ruby 94
Naselaris, Thomas 221
Nash, John 166
networks, human connectivity 178–86
neuroculture 210
neuroleadership 2
neuromarketing 210

neuroplasticity 6, 52–54
 and creativity 66
neuroscience, insights from 2–3
neurotechnology 220–25
neurotransmitters, link to emotions 107
Newberg, Andrew 207
Newell, Ben 135
Ney, Dr Julian 183
Nishimoto, Sinji 221
Nooyi, Indra 54
noradrenaline (norepinephrine) 107
Nordgren, Loran 134–35
Nowak, Professor Martin 172
nucleus accumbens 85

Oakley, Professor David 170
opiate reward centre 87
organizational routines 144–45
organizational structures, influence
 of neuroscience 223–24
organizational values 22
outlook (emotional style) 86–87
over-thinking, detrimental effects 142
oxytocin 187–88

panic mood 95, 96
Panksepp, Jaak 108–09
Parvizi, Josef 87
pattern recognition 153–54
 attraction for the brain 36–37
 potential traps 33–34, 36–37
performance focus 51–74
 adaptability in an unpredictable world 70–73
 creativity 62–66
 ever-changing brain 52–54
 flow state 59–62
 growth mindset 73
 impact of multitasking 26–30
 memory 67–70
 neuroplasticity 52–54
 purpose 54–59
 purposeful betting 71–73
 strategy in an uncertain world 71–73
persuasion 193–213
 brain-friendly communications 205–08
 changing behaviours 193–200
 compassionate communication 207–08
 influential words and phrases 205–08
 persuading the brain to act 194–200
 six principles of influence 200–05
 six stimuli model 210–13
pessimism–optimism continuum 86–87
Peters, Professor Steve 47
phase-locking 92–93

phronesis (practical judgement) 7
physical environment, influence on the
 unconscious mind 148–52
physical intelligence 151
physiology, interdependence with
 psychology 48
Pink, Daniel 55
Plutchik, Dr Robert 111–14, 121
positive psychology 116
Premack, David 172
priming the unconscious mind 136–43
 avoiding over-discussion 143
 avoiding over-thinking 142
 context of messaging 141
 in organizations 140–43
 influence of corporate culture 140
 power of words 141–42
 symbols 141
 use of logos 141
 verbal overshadowing effect 143
Prisoner's Dilemma, theory vs reality 164–69
psychology, interdependence with
 physiology 48
psychopathy 84
purpose
 altruism 57
 and leadership 56–57
 and motivation 54–59
 discovery 56
 excellence 56
 heroism 57
 higher purpose in organizations 56–57
 how to spot the wrong purpose 57–59
 intrinsic motivation 55–57
 'what' and 'how' questions 55
 'why' questions 55
purposeful betting 71–73

questioning
 anti-enquiring culture 41, 42, 43–44
 power of asking questions 41–43
 questioning styles 43–46
 evil interrogator 43
 good listener 43
 grumpy loner 43, 44
 star enquirer 43, 44

Ramachandran, V S 177
Random-Acts-of-Kindness (RAKs) 120–21
randomness 71
Rao, Professor Rajesh PN 220
Recognition-Primed Decision (RPD) model 153–54

Index

relations (Pillar 4) 10–11, 163–217, 216–17
   amygdala and trust 186–87
   communication 193–213
   cooperation 163–72
   human connectivity 178–86
   imitation 177–78
   machines replacing humans at work 188–89
   persuasion 193–213
   role of brain chemicals 187–88
   role of mirror neurons 177–78
   social network theory 179–83
   socially wired brain 169–72
   summary of Pillar 4 10–11, 216–17
   Theory of Mind (ToM) 172–76
   understanding how others are thinking 172–76
relationship management (external management) 99
relaxation techniques 48
Renvoise, Patrick 210–12
reptilian structures of the brain 35
resilience 86
return on investment (ROI) 72–73
Ridley, Matt 167–68
Rifkin, Jeremy 188
Robinson, Sir Ken 150
rocket mood 95, 96
Rodin, Auguste, *The Thinker* 34
Ronson, Jon 84
Rosen, Larry 148–49
Ruffini, Giulio 221–22

satisfaction, measurement of 117–18
Schmidt, Eric 27
self-awareness 88–91, 98
self-confidence 98
self-control
   and multitasking 28–29
   benefits of higher-level thinking 21
   benefits of immediate feedback 23
   use of brain energy 19–20
self-management (internal management) 98
Seligman, Martin 116
Sen, Amartya 165
serotonin 107
short-term memory 67
Simonides of Ceos 69
Sims, Peter 71
Sinek, Simon 55
six stimuli model of persuasion 210–13
small betting 71–73
social competence (external awareness) 98

social intuition 87–88
social media 183
social network theory 179–83
socially wired brain 169–72
sociopathy 84
Socrates 42
Stanford marshmallow experiment 19–20
stereotyping 39
strategy, in an uncertain world 71–73
strong values, and willpower 21–23
submission and obedience, dangers of 113
Sunstein, C R 166
supertasking phenomenon 30
survival, brain survival protocols 46–48
survival-obsessed brain 34–36
symbols, priming effect 141

Taleb, Nassim Nicholas 71
Taylor, Ros 62
teams and groupthink 37–38
TED speeches 55, 150, 183, 207
Tetlock, Philip 46
Thaler, Richard H 166
Theory of Mind (ToM) 172–76
thinking (Pillar 1) 8–9, 15–74, 76–77
   and evolution of the brain 34–36
   brain power 17–31
   clear state of mind 33–49
   performance focus 51–74
   summary of Pillar 1 8–9, 76–77
Thompson, Clive 148
Tomasello, Michael 173–74
Tomkins model of emotions 106–07, 108–11
total quality management (TQM) 117
transformational leadership style 7
trust, role of the amygdala 186–87

unconscious mind
   influence of the physical environment 148–52
   power of 131–36
Unconscious Thought Theory 134–35

values
   building organizational values 22
   strong values and willpower 21–23
van den Driest, Frank 223–24
vasopressin 187–88
verbal overshadowing effect 143
volatile, uncertain, complex and ambiguous (VUCA) world 5–6

Waldman, Mark 207
Watkins, Dr Alan 48

Weed, Keith 223–24
Wegner, Daniel 135
willpower
    and strong values 21–23
    benefit of immediate feedback 23
    benefits of higher-level thinking 21
    effects of ego depletion 20
    use of brain energy 19–21
    viewed as a muscle 19–21
    ways to strengthen the muscle 21–23

Wilson, Timothy 134
Wiseman, Professor Richard 120–21
Woodruff, Guy 172
words, priming power of 141–42
work ethic 21–22
Wright brothers 55

Yeung, Rob 205–06

Zimbardo, Philip 113, 140, 203